LED Lighting
The Third Revolution in Lighting

LED照明
第三次照明革命

主编 **欧阳东**
Editor-in-Chief **Ouyang Dong**

中国建筑工业出版社
China Architecture & Building Press

图书在版编目（CIP）数据

LED照明——第三次照明革命 / 欧阳东主编. —北京：中国建筑工业出版社，2015. 6
ISBN 978-7-112-17951-0

Ⅰ. ①L… Ⅱ. ①欧… Ⅲ. ①发光二极管－照明设计 Ⅳ. ①TN383.02

中国版本图书馆CIP数据核字（2015）第057545号

为贯彻执行国家技术经济和节能政策，推广宣传LED技术及相关LED照明产品，促进行业革命和进步而编写本书。全书共6章，内容包括：总则，LED照明技术，LED照明技术标准现状与发展规划，传统照明和LED照明的对比，LED照明技术的主要问题与对策，LED照明应用典型案例等。本书图文并茂、重点突出、中英文对照，有较强参考性、启发性、实用性。

本书既可供政府部门、建设单位、设计单位、施工单位、监理单位、照明厂商等单位的负责人、设计师、技术人员等学习参考，也可作为大中专学校相关专业师生及广大对LED照明感兴趣人士阅读。

责任编辑：刘　江　范业庶
装帧设计：北京美光设计制版有限公司
责任校对：张　颖　关　健

LED照明
——第三次照明革命
主编　欧阳东
*
中国建筑工业出版社出版、发行（北京西郊百万庄）
各地新华书店、建筑书店经销
北京美光设计制版有限公司制版
北京方嘉彩色印刷有限责任公司印刷
*
开本：880×1230毫米　1/32　印张：4 3/4　字数：173千字
2015年6月第一版　2015年6月第一次印刷
定价：49.00元
ISBN 978-7-112-17951-0
(27194)

编委会 Editorial Board

特邀编委 Guest Editorial Members

作者简介
About the Author

欧阳东

中国建筑设计研究院（集团）
院长助理、教授级高工
国务院政府特殊津贴专家
中国勘察设计协会建筑电气设计分会常务副理事长
中国建筑节能协会建筑电气节能专委会常务副主任
北京市和住房城乡建设部人才中心正高评委

Ouyang Dong

China Architecture Design & Research Group (CAG)
Assistant President, Professor-level Senior Engineer
Expert Entitled to Government Special Allowance (GSA)
Standing Vice President of China Intelligent Building Technology Information Association
Deputy Director of Building Electricity and Intelligent Efficiency Committee of China Association of Building Efficiency
Senior Judge of the Talent Center of the Beijing Municipality and the Ministry of Housing and Urban-Rural Development of the People's Republic of China.

欧阳东，1982年取得重庆建筑工程学院自动化专业的工学学士，2009年取得厦门大学EMBA高级经济管理硕士；2005年取得国家一级电气注册工程师。曾任综合所所长、机电院院长、运营中心主任等职务，现任集团院长助理、集团总法律顾问；社会兼职：《智能建筑电气技术》杂志社副社长；“中国智能建筑信息网”网站的理事长。

Mr. Ouyang Dong obtained his Bachelor's degree in Automation from Chongqing Institute of Architecture and Civil Engineering in 1982 and an Executive MBA (EMBA) in Senior Economics and Management from Xiamen University in 2009, and became a National First Class Registered Electrical Engineer in 2005. He was previously Director of the Comprehensive Department, President of the Electrical and Mechanical Institute, Director of the Operation Center of China Architecture Design & Research Group (CAG), and is currently Assistant to the CAG President and the Group General Counsel. His appointments include Deputy Director of the publication Intelligent Building Electrical Technology and President of the China Intelligent Building Information website.

作为工种负责人，参与了几十项大中型项目的设计工作，并取得了“北京梅地亚中心”等多个设计项目的国家级、省部级优秀设计奖。作为项目负责人，参与了多项企业级、部级和国家级科研项目，并取得了《建筑机电设备开放式通信协议研究》等多个科研项目的住房和城乡建设部华夏建设科学技术奖二、三等奖，主编了《医疗建筑电气设计规范》（中英文）；作为第一专利人，发明了《智能型灯光面板》等三项专利；作为主编或副主编完成了《建筑机电节能设计手册》等十几本著作的编著工作，并均已正式出版；独著《建筑机电节能设计探讨》、《设计企业管理研究》等十几篇技术论文和管理论文。主持过多次全国性行业会议，多次在行业会议上宣讲了《建筑机电节能设计研究》、《管理创新——企业发展之精髓》、《BIM技术——建筑设计的第二次革命》。

As principal of his profession, Mr. Ouyang has participated in designing dozens of large and medium-sized projects, and several of his design projects, such as Media Center Hotel Beijing have earned national and provincial excellent design awards. As project principal, he has been involved in several corporate, ministerial and national scientific research projects. Mr. Ouyang's multiple scientific research projects like Open Communication Protocol Study of Building Mechanical and Electrical Equipment were awarded second and third prizes in the China Construction Science and Technology Awards from the Ministry of Housing and Urban-Rural Development of the People's Republic of China and he also edited Code for the Electrical Design of Medical Buildings (Chinese and English versions). In addition, Mr. Ouyang holds three patents as the first inventor, including one for Intelligent Type lamplight Switch Panel. As chief or associate editor, he has completed the compilation of a dozen works that have been published, like Building Mechanical and Electrical Efficiency Design Manual, and a dozen of academic papers in technology and management, such as Discussion on Building Mechanical and Electrical Efficiency Design and Study on Design Enterprise Management. Moreover, he has chaired national industry conferences numerous times and presented papers on Study on Building Mechanical and Electrical Efficiency Design, Management Innovation: the Essence of Corporate Development, and BIM Technology: The Second Revolution in Architectural Design.

作为院（集团）院长助理兼设计运营中心主任，组织完善、调整、建立了一套《新的设计组织架构体系——项目经理和设计研究室主任的强矩阵管理架构》，取得非常好的经营业绩，各项经营指标连续四年均创历史新高。作为负责人组织BIM技术应用和推广工作，并取得了BIM最佳企业应用奖和五个BIM项目最佳设计奖。曾获得院（集团）管理创新特殊贡献奖、“十一五”科技创新奖、科研管理奖。

As Assistant President of CAG and Director of Design Operation Center, Mr. Ouyang has conducted, adjusted and established a set called New Design Organizational Structural System: Strong Matrix Management Structure of the Project Manager and Design & Research Office Director, and yielded outstanding business results with all business indicators hitting the record highs for four consecutive years. As principal, he has implemented the application and promotion of BIM technology, and was awarded the Best Enterprise with BIM Application and five Awards of the Best Design for BIM Projects. He was also awarded the CAG Special Services Award to Management Innovation, and 11th Five-Year Technology Innovation Award and Scientific Research Management Award.

序
Foreword

随着社会日新月异的发展，半导体照明在全球范围内都是朝阳的新兴产业。半导体照明具有节能减排、寿命长、体积小的特点，横跨了传统照明产业、电子行业等多个领域，具有广阔的发展前景，并符合“国家‘十二五’中长期科技发展规划战略研究”的发展方向。

在照明行业中，LED照明作为第三代照明新型光源已逐渐被采用，并替代原有的第一代或第二代照明光源。LED照明更是一个涉及多个领域、综合性很强的产业，它包含的大量信息，需要强有力的技术手段去采集、分类、分析、检索和传输；而建筑信息模型BIM（Building Information Modeling）技术作为数字建筑技术中出现的新概念、新理念和新技术，将为建筑设计革命提供强有力的技术支撑。2014年10月7日，日本科学家赤崎勇（Isamu Akasaki）、天野浩(Hiroshi Amano)和美籍日裔科学家中村修二(Shuji Nakamura)因发明了蓝色发光二极管以及基于此技术的白光LED照明应用，极大地促进了新一代高效节能的照明革命，并因此荣获了2014年诺贝尔物理学奖。

目前，国家通过政策扶持、资金扶持等多种手段，大力推广LED照明在照明领域的应用。中国为了更好地掌控LED产业，也有了自己的技术和产业发展思路：抓住照明产业革命的历史机

As the society develops rapidly, with the features of energy-saving, emission reduction, long life as well as small size, LED lighting is becoming a promising emerging industry at the global level. It spans the fields of traditional lighting industry and electronics industry with vast development prospects. Moreover, it accords with the development direction in the ‘the national 12th Five-year medium and long term science and technology development plan’.

In the lighting industry, LED lighting has been gradually used as the third generation new light source replacing the 1st and 2nd generations light sources. Being comprehensive across many areas, Led lighting includes plenty of information which requires powerful technical to gather, classify, analyze, search and transmit. At the same time, as the emerging new concept, new philosophy and new technique in the digital building technology, Building Information Modeling Technology is providing powerful technological support for architectural design revolution. On the 7th of October, 2014, Japanese scientists IsamuAkasaki and HiroshiAmano and Japanese-American scientist ShujiNakamura were awarded the Nobel Prize for physics for their invention of blue light emitting diodes and the white light LED lighting, which greatly promotes the new revolution of high-efficiency and energy-saving lighting.

At present, China is vigorously promoting the LED lighting application in the lighting field by means of policies support and Financial Support. In order to have a better development of LED industry, China has its own technology and industry development ideas. Seize the historical opportunity of the lighting industry revolution, and adhere to the government guide; Take enterprises as the subjects and marketization operation as the principle, and take technology innovation as the core and mechanism

遇，坚持政府引导；以企业为主体和市场化运作原则，以技术创新为核心、机制创新为保障；在解决市场继续的产业化技术的同时，加大对重大关键技术的研发投入，集中力量，重点突破，实现跨越式发展；通过全球范围内资源的整合、基地建设和龙头企业的培育，形成有自主知识产权和有国际竞争力的新兴产业。

在全球经济化、市场化、知识化、信息化的今天，也是跨行业、跨专业、跨产品的跨界社会，在这样一个挑战和机遇并存的环境下，中国的LED产业必将迎来一个崭新的未来。

innovation as the security; Increase investment into the major key technology research and development while solving the problem of industrialization technology. Concentrate and break through at key points and realize leap-forward development. Through the integration of global resources, base construction and the cultivation of leading enterprises, make LED lighting an emerging industry of Independent intellectual property rights and international competitiveness.

Under global economization, marketization, knowledge-driven and informatization and a cross-industry, cross-discipline and cross-product society presenting both opportunities and challenges, China LED industry is having a new future.

国家半导体照明工程研发及产业联盟执行主席

中国科学院半导体照明研发中心主任

Li Jinmin

China Solid State Lighting Alliance 2015.2.6 Executive Chairman

Research and Development Center for Semiconductor Lighting Chinese Academy of Sciences Director

Feb 6.2015

前言
Preface

形势 Trend	中国在“十一五”时期取得了辉煌业绩，国民生产总值（GDP）从26.6万亿元增加到51.9万亿元，年增速9.3%，跃升到世界第二位；5年科技累计投入8729亿元，年增速超过18%；在科技创新和节能环保的大背景下，LED 产业作为节能环保的重要产业之一；预计到2015年中国LED照明产业规模将达到5000亿元；因此，节能减排和低碳经济将是中国发展的永恒主题。 China obtained brilliant achievements during the 11th five-year plan in which China’s GDP had an annual growth rate of 9.3%, an increase from 26.6 trillion Yuan to 51.9 trillion Yuan and the economy ranked second in the world. Investment in sciences and technology amounted to 872.9 billion Yuan in five years with an annual growth rate of more than 18%. Meanwhile, science and technology innovation and energy saving and environmental protection were proposed and LED industry should be an important industry of energy saving and environmental protection. The LED lighting industry scale was predicted to be 500 billion Yuan in 2015. Therefore, energy efficiency and emission reduction and low carbon economy are going to be the perpetual trend of development in China.
目的 Goal	在中国建筑行业，通过LED照明技术和国家补贴政策等方式，推广LED技术，促进行业科技进步。贯彻执行国家技术经济和节能政策，推广宣传LED照明产品，促进行业革命和进步，将LED产品大力推向市场，为国家的节能减排尽力。 In the construction industry in China, popularize LED technology and promote the industrial technical progress by way of LED lighting technology and national allowance policies. Implement national technical, economic and energy saving policies, generalize and publicize LED lighting products, accelerate industrial revolution and advancement as well as bring LED products onto market, so as to contribute to the energy saving and emission reduction of our state.

（续表）（Continued）

主题 Theme	2011年，科技部颁布《关于印发国家十二五科学和技术发展规划的通知》，节能环保位居七大战略性新兴产业之首，其中，LED照明又位居四大节能环保技术之首。国家将低碳经济作为带动经济增长的内生增长动力；半导体照明是中国在节能减排方面的成功突破口；光电子技术与微电子技术的融合，不仅带来光的革命，更带来高新技术革命。 In 2011, Ministry of Science and Technology issued 'the notice about printing and distributing the national 12th Five-year science and technology development plan', in which the industry of energy saving and environmental protection topped the seven strategic emerging industries, among which LED lighting topped the four technologies of energy saving and environmental protection technologies. The state regards the Low carbon economy as the endogenous growth force driving economic growth. Semiconductor lighting is the successful breakthrough of china in energy saving and emission reduction. The merging of optoelectronic and microelectronic technologies not only brings the revolution of light but also brings the revolution of high-tech.
需求 Needs	1.经济发展的需求（GDP、生产力、可持续健康发展等）； 2.技术进步的需求（手段、流程、质量、效率等）； 3.行业发展的需求（科技研发、节约投资等）； 4.社会进步的需求（节能环保、社会责任等）。 1.Economic development needs (GDP, productivity, sustainable and healthy development , etc.) 2.Technological progress needs (means, processes, quality, and efficiency, etc.) 3.Industrial development needs (science and technology R&D and investment savings, etc.) 4.Social progress needs (energy saving and environmental protection, and social responsibility, etc.)

（续表）（Continued）

内容 Content	总则，LED照明技术，LED照明技术标准现状与发展规划，传统照明和LED照明的对比，LED照明技术的主要问题与对策，LED节能改造案例等。 General, LED Lighting Technology, Current Situation & Development Planning of LED Lighting Technical Standards, Comparison between Traditional Lighting & LED Lighting, Main Problems & Countermeasures of LED Lighting Technology, LED-Based Energy-Saving Renovation Cases, etc.
对象 Object	政府部门、建设单位、设计单位、施工单位、监理单位、照明厂商等单位的技术人员等。 Technical personnel of governments, project owners, design units, construction units, supervision units and Lighting manufacturers.
特点 Features	图文并茂、突出重点、可借鉴性、中英文对照、参考性、启发性、实用性。 Rich images and texts, highlighting key points, adoptability, bilingual in Chinese and English, providing reference, enlightening and practicability.

（续表）(Continued)

不足 Inadequacies	由于大家都是利用业余时间，在短时间内的编制完成，采用了大量的国内外相关资料，翻译经验不足，若有不妥或不准确之处，请大家批评指正。 Inappropriateness and inaccuracy are inevitable due to the short compilation time as the editors wrote this book in spare time, the massive amounts of domestic and overseas relevant data used as well as the inadequate translation experience. Constructive feedbacks are welcome.

中国建筑设计研究院（集团）院长助理、
教授级高工、国务院政府特殊津贴专家
2015-2-6
Ouyang Dong
China Architecture Design & Research Institute (Group)
Assistant to the President, Professor-level Senior Engineer
Expert Entitled to Government Special Allowance (GSA)
Feb 6, 2015

目录
Contents

本书重点摘要
Summary

1 总则
General

三次照明革命对照比较

Contrast of The Three Lighting Revolutions

时代 Age	第一次照明革命（1879年） The 1st Lighting Revolution (1879)	第二次照明革命（1938年） The 2nd Lighting Revolution (1938)	第三次照明革命（1962年） The 3rd Lighting Revolution (1962)
名称 Name	白炽灯 Incandescent Lamp	荧光灯 Fluorescent Lamp	LED
发明者 Inventor	1879年，美国科学家爱迪生（1847~1931）32岁发明了白炽灯。 In 1879, American inventor Thomas Edison（1847~1931）invented incandescent lamp when he was 32 years old.	1938年4月1日，美国通用电子公司伊曼（1895~1972）43岁发明了荧光灯（日光灯）。 On April 1, 1938, George Inman（1895~1972）of General Electric invented fluorescent lamp when he was 43 years old.	1962年，美国通用电气公司（GE）的Nick Holonyak Jr博士(1928~今）34岁发明了可见光的LED。 In 1962, Dr. Nick Holonyak Jr. of GE invented LED that gives out visible light when he was 34 years old. 2014年诺贝尔物理学奖联合授予日本科学家赤崎勇、天野浩以及美籍日裔科学家中村修二，表彰他们发明一种新型高效节能光源[蓝色发光二极管（LED）]。 In 2014, Nobel Prize in Physics was awarded to Isamu Akasaki, hiroshi Amano and Shuji Nakamura for their new invention of efficient blue Light-emitting diodes (LED).

（续表）（Continued）

时代 Age	第一次照明革命（1879年） The 1st Lighting Revolution (1879)	第二次照明革命（1938年） The 2nd Lighting Revolution (1938)	第三次照明革命（1962年） The 3rd Lighting Revolution (1962)
技术特点 Technical Features	用钨丝加热发光，属于热辐射光源。发光效率10~15lm/W。 Giving light by heating tungsten filaments the lamp is a heat radiation light source. Luminous efficiency: 10~15lm/W.	用汞蒸气和荧光粉发光，属于气体放电光源。发光效率60~90lm/W。 Giving light by utilizing mercury-vapor and fluoresence, fluorescent lamp is a gas discharge light. Luminous efficiency: 60~90 lm/W.	用半导体发光，属于固态光源。当前主流产品发光效率80~120lm/W，理论发光效率 可达350lm/W。 LEDS use semiconductor to give light and belong to solid-state light. At present, the luminous efficiency of mainstream LEDS is 80~120lm/W. LED'S theoretic luminous efficiency can be 350lm/W.
优点 Strength	价格便宜，安装方便，技术成熟。 Low price, easy to install, mature technology.	价格合理，节能较大，发光效率适中。 Reasonable price, good energy-saving, medium luminous efficiency.	体积小、耗电低、寿命长、无毒环保，易于智能控制。 Small size, low power consumption, long life, non-toxic and green, and smartly controlled.
缺点 Weakness	光效低，发热量大，寿命短。 Low luminous efficiency, high heating value,short lifetime.	寿命一般。 Medium lifetime.	散热技术；发光驱动电路。 Heat dissipation technology and lighting drive circuit.

1.1 LED照明设计理念
Design Concepts of LED Lighting

功能需求：根据建筑功能，按照设计标准及规范，实现照明功能的要求。

Functional demand: Light up a building according to the building functions and the design standards and codes.

节能需求：在满足同等舒适度的条件下，实现照明节能。

Energy-Saving demand: Use less energy while light up a room and offer the same comfortability.

舒适需求：满足人的生理、心理要求，提高环境的光品质。

Comfortability demand: Fulfill people's physiological and psychological demands and light up the environment comfortably.

文化需求：营造光环境，提升建筑空间的艺术效果。

Cultural demand: Create better lighting to enhance the art effects of an architectural space.

1.2 LED照明产业应用
Applications of LED Lighting Industry

以LED为主线，产业上中下游覆盖诸多领域。

Concentrated on LED, the whole industry covers multiple fields.

LED通用照明市场渗透率在2015年将达到30%以上，白光发光二极管的发光效率达到国际同期先进水平，推动我国半导体照明产业进入世界前三强。

LED will have more than 30% of the shares in general lighting market in 2015. The luminous efficiency of white LEDS in China will be as advanced as that in the world. China's LED industry will rank top three in the world.

1.3 LED照明产业流程
Flow of LED Lighting Industry

上游产业 Upstream → 材料、外延、芯片 Mateirals, extension & chips

中游产业 Midstream → 封装、模组 Packaging & modules

下游产业 Downstream → 应用、系统 Applications & systems

1.4 LED照明基本要求
Basic Requirements of LED Lighting

视觉舒适： Visual Comfort:	光源的显色性、和谐亮度分布 Color Rendering of Photo Source and Harmonious Brightness Distribution
视觉质量： Visual Performance:	照度水平和眩光控制 Lighting Level and Glare Limitation
视觉气氛： Visual Ambience:	光源色温、光的方向和光的阴影 Lighting Color, Direction of Light and Modeling

1.5 LED照明产业现状
Current Situations of LED Lighting Industry

上游：LED芯片和封装材料企业，主要在珠三角、长三角、环渤海等地区；

Upstream: LED Chip enterprises and Packaging material Enterprises, most of which lie in Pearl River Delta, Yangtze River Delta and Bohai Rim.

下游：LED灯具企业，主要在珠三角、长三角地区。

Downstream: LED Lamp enterprises, most of which lie in Pearl River Delta and Yangtze River Delta Regions.

LED灯比白炽灯节电90%，LED灯比节能灯节电30%。

LED Light consume 90% less electricity than incandescent lamps do and 30% less electricity than energy saving lamps do.

为推动LED技术在我国建筑设计市场的应用发展，承担住房城乡建设部科研课题《中国建筑电气与智能化节能发展报告》，编写其中第六章。

In order to drive the application and development of led technology in China's architectural design market, we Undertake the Scientific Research Project of MOHURD Report on the Development of Electric and Intelligent Energy-saving in China Buildings and write Chapter 6 of it.

1.6 LED照明产业发展趋势
Development Trends of LED Lighting Industry

发展趋势之一：政策扶持
发展趋势之二：资金支持
发展趋势之三：技术发展
发展趋势之四：产品发展——替代传统照明
发展趋势之五：市场发展——LED照明市场前景

Trend 1: Policies Support
Trend 2: Financial Support
Trend 3: Technological Development
Trend 4: Product Development——Replacement of the Traditional Lighting
Trend 5: Market Development——Outlook of LED lighting market

1.7 未来LED照明创新技术
LED Lighting Innovation Technologies in Future

创新技术之一：解决飞机时差
创新技术之二：解决集中精力
创新技术之三：解决睡觉失眠
创新技术之四：解决果蔬保鲜
创新技术之五：促进植物生长
创新技术之六：解决交通站牌
创新技术之七：解决产品形态
创新技术之八：解决健康照明（舒适）
创新技术之九：解决智能控制（系统）
创新技术之十：解决可见光通信

Innovation Technology 1：Solve Jet Lag
Innovation Technology 2：Promote Concentration
Innovation Technology 3：Alleviate Insomnia
Innovation Technology 4：Fruits and vegetables Fresh-keeping
Innovation Technology 5：Promote Plants Growth
Innovation Technology 6：Transport Station Boards
Innovation Technology 7：Diversify Product Forms
Innovation Technology 8：Healthy Lighting (Comfort)
Innovation Technology 9：Intelligent Control (System)
Innovation Technology 10：Visible Light Communication

2 LED照明技术
LED Lighting Technology

2.1 LED照明技术特点
Technical Features of LED Lighting

技术特点之一：衬底技术
技术特点之二：外延技术
技术特点之三：芯片技术
技术特点之四：封装技术
技术特点之五：散热器技术
技术特点之六：驱动技术
技术特点之七：配光技术

Feature 1: Substrate Technique
Feature 2: Wafer Technique
Feature 3: Chip Technique
Feature 4: Packaging Technique
Feature 5: Radiator Technique
Feature 6: Drive Technique
Feature 7: Light Distribution Technique

2.2 LED照明节能特性
Energy Saving Features of LED Lighting

综合节能特性之一：光效高
综合节能特性之二：能耗低
综合节能特性之三：寿命长
综合节能特性之四：易控制
综合节能特性之五：低运维
综合节能特性之六：安全环保
综合节能特性之七：光谱丰富
综合节能特性之八：绿色环保

Energy Saving Feature 1: High Luminous Efficiency
Energy Saving Feature 2: Low Energy Consumption
Energy Saving Feature 3: Long Life
Energy Saving Feature 4: Easy Control
Energy Saving Feature 5: Easy Operation & Maintenance
Energy Saving Feature 6: Safe and Clean
Energy Saving Feature 7: Broad Spectum
Energy Saving Feature 8: Green & Environment-Friendly

3 LED照明技术标准现状与发展规划
Current Situations and Development Plan of LED Lighting Technical Standards

3.1 国外LED照明技术标准现状
Current Situations of Foreign LED Lighting Technical Standards

欧洲LED标准 LED Standards in Europe	北美LED标准 LED Standards in North America	亚洲LED标准 LED Standards in Asia

3.2 中国LED照明技术标准现状
Current Situation of LED Lighting Technical Standards in China

3.3 中国LED照明技术相关政策
Policies related to LED Lighting Technologies in China

3.4 LED照明技术发展规划
Development Plan of LED Lighting Technologies

4 传统照明和LED照明的对比
Traditional Lighting VS LED Lighting

对比之一：光源性质关系
Difference 1: Properties of Photo Sources
对比结果：LED安全，无污染，光电转换效率高,采用LED是必然的。
Conclusion: Due to the safety, pollution-free and high photoelectric conversion efficiency of LED, it is inevitable to employ LED.

对比之二：节能降耗
Difference 2: Energy-Saving and Consumption Reduction
对比结果：LED发光效率高,寿命长，运维费低，是未来发展趋势。
Conclusion: High luminous efficiency, long life and low operation and maintenance cost of LED make it the development trend in future.

对比之三：光品质比较

Difference 3: Quality of Light

对比结果：LED光源的色温可变，光稳定性好，应用广泛。

Conclusion: Due to the variable color temperature and high photostability of LED light source, it is widely used.

对比之四：驱动控制比较

Difference 4: Drive Control

对比结果：LED光源稳态照明时间短，待解决驱动成本问题后，更具有竞争力。

Conclusion: With short steady-state lighting time, LED light source will be more competitive once drive cost is reduced.

对比之五：设计及应用比较

Difference 5: Design and Application

对比结果：LED可一体化设计，光源灵活多变，光色可变，应用为必然趋势。

Conclusion: Due to the possible integrated design, flexible and variable light source and variable light color of LED, it is inevitable to employ LED.

对比之六：光源形状比较

Difference 6: Shapes of Photo Sources

对比结果：因LED光源的形状可变性，透光材料任选性，并可二次设计，所以LED广泛应用。

Conclusion: Due to the variable light source shape, optional light-transparent material and possible secondary design of LED, it is widely used.

对比之七：可见光通信比较

Difference 7: Visible Light Communication

对比结果：由于LED有可见光通信，信息承载能力高，实现可见光通信成为可能。

Conclusion: Because of the visible light communication and high information bearer capability of LED, it is possible to realize visible light communication.

对比之八：光源的综合成本比较

Difference 8: Composite Cost of Light Sources

对比结果：因为LED综合成本相对较低，所以LED应用是发展方向。

Conclusion: The low composite cost of LED makes it the development trend.

对比之九：非可视照明应用比较

Difference 9: Nonvisual Lighting

对比结果：从LED的健康医疗、种植养殖、工艺用光三个方面的扩展应用上看，是未来的发展方向。

Conclusion: The extended applications of LED in healthcare, planting & breeding and technical lighting are going to be the development trends in future.

5 LED照明技术的主要问题与对策

Main Problems & Countermeasures of LED Lighting Technology

5.1 LED照明技术的主要问题

Main Problems of LED Lighting Technology

问题之一：可靠性
问题之二：光生物安全
问题之三：舒适性
问题之四：成本
问题之五：光效
问题之六：标准及规范
问题之七：照明形态
问题之八：智能化

Problem 1: Reliability
Problem 2: Photobiological Safety
Problem 3: Comfort
Problem 4: Cost
Problem 5: Luminous Efficiency
Problem 6: Standards and Norms
Problem 7: Lighting Forms
Problem 8: Intelligence

5.2 LED照明技术的主要对策

Countermeasures for LED Lighting Technology

对策之一：政策的扶持
对策之二：技术的提升
对策之三：商业模式创新
对策之四：金融扶持
对策之五：LED标准的制定

Countermeasure 1: Policies Support
Countermeasure 2: Technical Advance
Countermeasure 3: Business Mode Innovation
Countermeasure 4: Financial Support
Countermeasure 5: LED Standards Development

6 LED节能改造案例

LED-Based Energy-Saving Renovation Cases

总则

General

三次照明革命对照比较

Contrast of the Three Lighting Revolutions

时代 Age	名称 Name	发明者 Inventor	技术特点 Technical Features	优点 Strength	缺点 Weak-ness
第一次照明革命（1879年）The 1st Lighting Revolution (1879)	白炽灯 Incandescent lamp	1879年，美国科学家爱迪生（1847~1931）32岁发明了白炽灯。In 1879, American inventor Thomas Edison (1847~1931) invented incandescent lamp when he was 32 years old.	用钨丝加热发光，属于热辐射光源。Giving light by heating tungsten filaments the lamp is a heat radiation light source. 发光效率10~15lm/W luminous efficiency: 10~15lm/W	价格便宜，安装方便，技术成熟。Cheap, easy installation and mature technology.	光效低，发热量大，寿命短。Low luminous efficiency, high heating value and short lifetime.
第二次照明革命（1938年）The 2nd Lighting Revolution (1938)	荧光灯 Fluorescent lamp	1938年4月1日，美国通用电子公司伊曼（1895~1972）43岁发明了荧光灯（日光灯）。On April 1, 1938, George Inman (1895~1972) of General Electric invented fluorescent lamp when he was 43 years old.	用汞蒸气和荧光粉发光，属于气体放电光源。Giving light by utilizing mercury-vapor and fluorescence, fluorescent lamp is a gas discharge light. 发光效率60~90lm/W luminous efficiency: 60~90lm/W	价格合理，节能较大，发光效率适中。Reasonable price, energy-saving and mediocre luminous efficiency.	寿命一般。Average lifetime.

（续表）（Continued）

时代 Age	名称 Name	发明者 Inventor	技术特点 Technical Features	优点 Strength	缺点 Weak-ness
第三次照明革命（1962年）The 3rd Lighting Revolution (1962)	LED	1962年，美国通用电气公司（GE）的Nick Holonyak Jr博士（1928~）34岁发明了可见光的LED。In 1962 Dr. Nick Holonyak Jr. (1928~) of GE invented LED that gives out visible light when he was 34 years old. 2014年诺贝尔物理学奖联合授予日本科学家赤崎勇(IsamuAkasaki)，天野浩(HiroshiAmano)以及美籍日裔科学家中村修二(ShujiNakamura)，表彰他们发明一种新型高效节能光源（蓝色发光二极管(LED)）In 2014, Nobel Prize in Physics was awarded to Isamu Akasaki, Hiroshi Amano and Shuji Nakamura for their new invention of efficient blue light-emitting diodes (LED).	用半导体发光，属于固态光源。LEDs use semiconductor to give light, and belong to solid-state light. 当前主流产品发光效率80~120lm/W，理论发光效率可达350lm/W。At present, the luminous efficiency of mainstream LEDs is 80~120lm/W. LED's theoretic luminous efficiency can be 350lm/W.	体积小、耗电低、寿命长、无毒环保，易于智能控制。Small size, low power consumption, long lifetime, toxic free, environment friendly and smartly controlled.	散热技术；发光驱动电路。Heat dissipation technology; drive circuit for lighting.

1.1 LED照明设计理念

Design Concepts of LED Lighting

序号 SN	需求 Demands	内容 Content
1	功能需求 Functional demand	根据建筑功能，按照设计标准及规范，实现照明功能的要求。 Light up a building according to the building functions and the design standards and codes.
2	节能需求 Energy-saving demand	在满足同等舒适度的条件下，实现照明节能。 Use less energy while light up a room and offer the same comfortability.
3	舒适需求 Comfortability demand	满足人的生理、心理要求，提高环境的光品质。 Fulfill people's physiological and psychological demands and light up the environment comfortably.
4	文化需求 Cultural demand	营造光环境，提升建筑空间的艺术效果。 Create better lighting to enhance the art effects of an architectural space.

1.2 LED照明产业应用

Applications of LED Lighting Industry

（1）以LED为主线，产业上中下游覆盖诸多领域。

（2）LED通用照明市场渗透率在2015年将达到30%以上，中国白光发光二极管的发光效率达到国际同期先进水平，推动中国半导体照明产业进入世界前三强。

Concentrated on LED, the whole industry covers multiple fields.

LED will have more than 30% of the shares in general lighting market in 2015. The luminous efficiency of white LEDs in China will be as advanced as that in the world. China's LED industry will rank top three in the world.

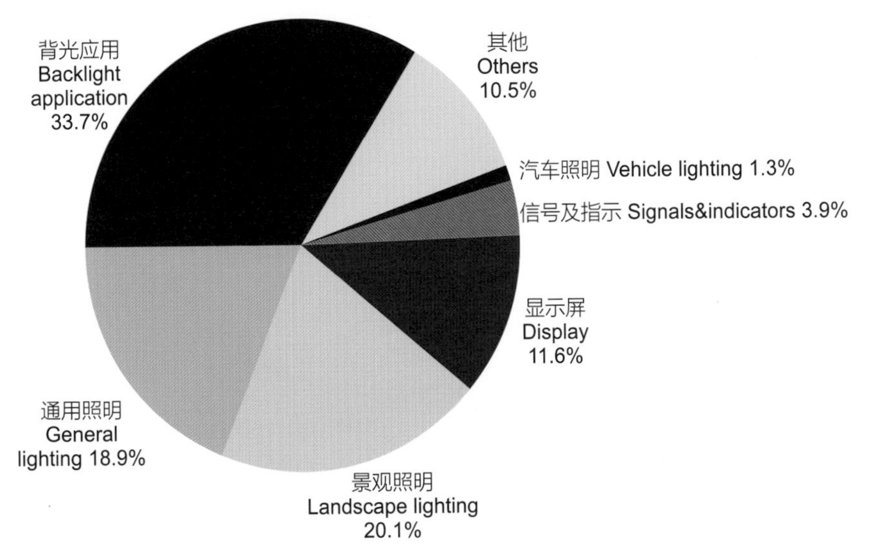

2013年中国LED照明应用领域分布
LED Lighting Applications in China in 2013

1.3 LED照明产业流程
Flow of LED Lighting Industry

1. 产业链
Industry Chain

上游产业 Upstream	中游产业 Midstream	下游产业 Downstream
材料、外延、芯片 Materials, extension & chips	封装、模组 Packaging & modules	应用、系统 Applications & systems

2. 产业生态环境

Ecological Environment of the Industry

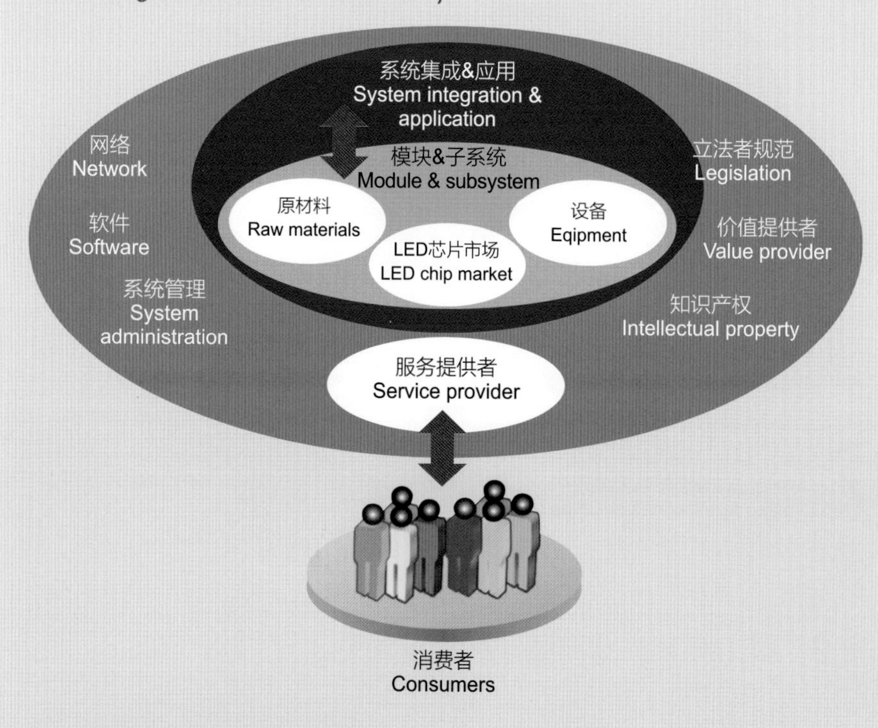

3. 基础研究与应用研究同步

Keep basic research and applied research at the same pace

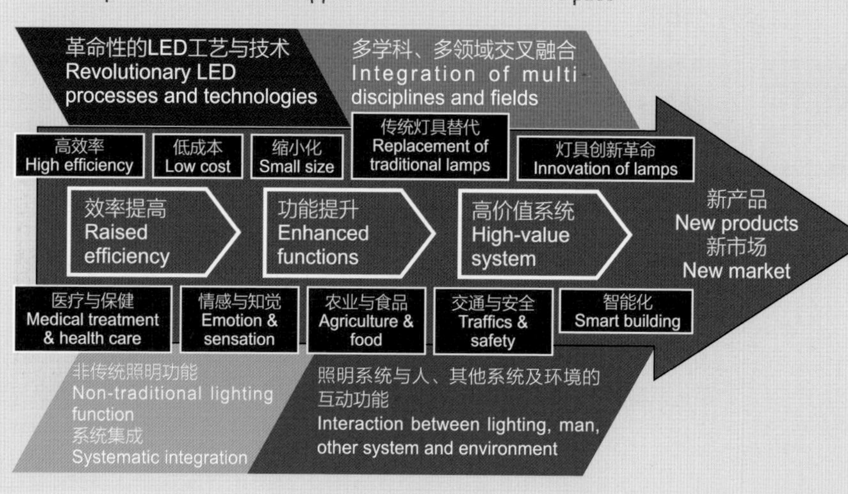

国家三大重点计划（LED相关内容）
Three Key Plans of the Government (LED related)

"973"计划 "973" Plan	"863"计划 "863" Plan	支撑计划 Support Plan
主题：高效氮化物材料研究 Subject: Research on efficient nitride materials 定位：探索性基础研究 Definition: Exploratory basic research	主题：关键材料、器件、装备和核心专利研究 Subject: Research on key materials, parts, equipment and core patents 定位：全面布局，重点突破 Definition: Overall planning and breakthroughs at key points	主题：应用产品开发、系统集成 Subject: Development & system integration of applied products 定位：示范推广 Definition: Modeling and promotion

1.4 LED照明基本要求

Basic Requirements of LED Lighting

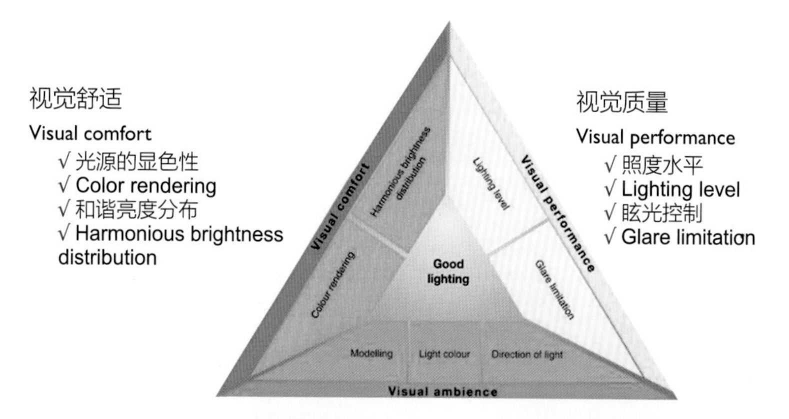

视觉舒适

Visual comfort

√ 光源的显色性
√ Color rendering
√ 和谐亮度分布
√ Harmonious brightness distribution

视觉质量

Visual performance

√ 照度水平
√ Lighting level
√ 眩光控制
√ Glare limitation

视觉气氛

Visual ambiance

√ 光源色温
√ Lighting color
√ 光的方向
√ Direction of light
√ 光的阴影
√ Modeling

现代照明的要求——只有LED照明使三个方面完美结合

Requirements for modern lighting —— Only LED lighting can fulfill the three demands.

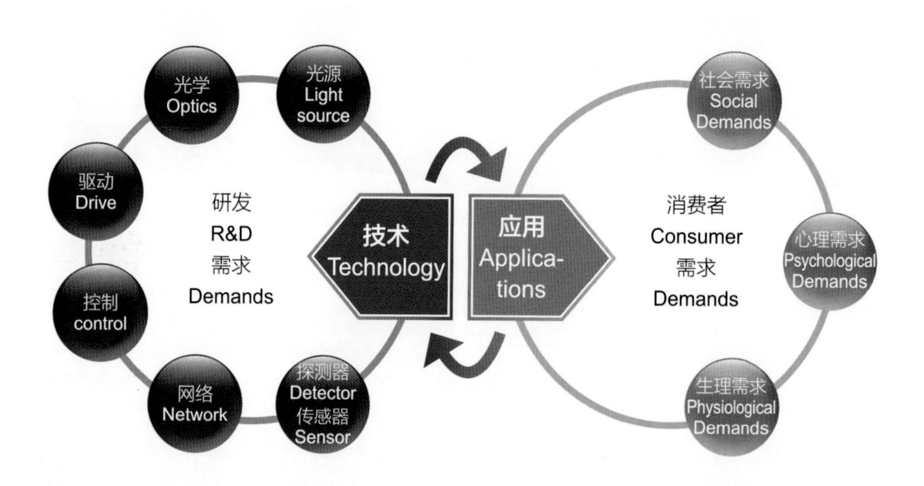

LED照明设计要求
Requirements of LED Lighting

根据《建筑照明设计标准》GB50034-2013的规定，并结合相关产品标准，参照有关国际标准要求，对LED灯提出以下技术要求：

Pursuant to Standard for Lighting Design of Buildings GB50034-2013, related product standards and international standards, the technical requirements of LED lamps are listed below:

（1）同一场所选用同类光源的色容差不应大于5SDCM。

Color tolerance adjustment of the same type of light source should not be more than 5SDCM at a same place.

（2）LED光源的“光通维持率”应符合规定要求。

LED light source's "lumen maintenance rating" should conform to related stipulations.

（3）在寿命期内LED光源的色品坐标与初始值的偏差在CIE1976均匀色度标尺图中，不应超过0.007，称作“色维持”。

In its lifetime the difference between LED light source's chromaticity coordinates and initial value should not be more than 0.007 in CIE 1976 Chromaticity Diagram, which is called "color maintenance".

（4）人长时间工作或停留的场所，LED光源的显色指数（Ra）不应小于80。

At places where men work or stay for a long time, LED light source's general color rendering index (Ra) should not be less than 80.

LED灯的色温不宜高于4000K，LED灯的特殊显色指数R9（饱和红色）应大于0。

LED lamp's color temperature should not be higher than 4000K, and special color rendering index R9 (deep red) should be more than 0.

（5）LED光源表面亮度高，容易导致眩光大和不舒适，应适当降低表面亮度（如不大于100kcd/m²）；灯具应限制眩光，宜有漫射罩或有30°的遮光角。

LEDs are very bright so that they will cause glare and discomfort to people. The brightness should be reduced (like not more than 100kcd/m²). Measures should be taken to control glare. It is best that LEDs work with diffusion cover or be designed into a 30° shielding angle.

（6）LED灯具在不同方向上的色品坐标，与其加权平均值偏差在CIE1976均匀色度标尺图中，不应超过0.004，即空间色品一致性。

The difference between LED lamp's chromaticity coordinates in different directions and its weighted mean should not be more than 0.004 in CIE 1976 Chromaticity Diagram, which means the spatial chromaticity should be consistent.

（7）应限制LED光源及其驱动电源的高次谐波（特别是3次谐波）含

LED light and its driver power's high-order harmonic (especially the third harmonic)

量，应符合《电磁兼容限值谐波电流发射限值（设备每相输入电流≤16A）》GB17625.1-2012中C类（照明设备）规定的谐波限值。

should be restricted and conform to the limits stipulated for Class C (lighting equipment) in Electromagnetic Compatibility–Limits–Limits for Harmonic Current Emissions (Equipment Input Current≤16 APer Phase) GB 17625.1-2012.

（8）应对LED输入电路的功率因数提出要求：LED光源及电源的功率因数（λ）低，是由于谐波过大所导致，这和电感镇流器和感应电动机是由于无功所引起的不同，所以从根本上说是要降低谐波含量。

Power factor of LED input circuit should be under restriction: LED light and power's low (λ) power factor is due to excessive harmonics. It is totally different from the low power factor occurred in inductor ballast and induction motor, which is caused by reactive power. Fundamentally, harmonics should be reduced.

《普通照明用非定向自镇流LED灯性能要求》GB/T 24908-2014（征求意见稿）灯的颜色性能

LED's colors stipulated in GB/T 24908-2014 Non-directional Self-ballasted LED-lamps for General Lighting Services -- Performance Requirements" (Draft)

色调规格 Tones	色调代码 Tone codes	一般显色指数Ra General color rendering index Ra	色坐标目标值 Target values in color coordinate		色容差SDCM Color tolerance adjustment SDCM	颜色不均匀度 Color non-uniformity	颜色漂移（色维持） Color shift (color maintenance)
			X	Y			
6500K（日光色）（RR） 6500K (daylight) (RR)	65	80（标称高显色指数的90） 80 (nominal high CRI at 90)	0.313	0.337	≤5	CIE1976（u'，v'）图上，灯在光束角范围内各方向上颜色坐标与平均值偏差$\Delta u'$、v'应≤0.005 In CIE 1976 (u' and v') chromaticity diagram the difference $\Delta u'$ and v' between color coordinates and mean value in every direction within the scope of beam angle should be ≤0.005.	点燃3000h平均色坐标相对于初始值坐标漂移应$\Delta u'$、v'≤0.005；6000h时应≤0.007 When a LED is lighted up for 3000h, the shift of mean color coordinates relative to initial coordinates $\Delta u'$ and v' should be ≤0.005; When it is lighted up for 6000h, they should be≤0.007.
5000K（中性白色）（RZ） 5000K (neutral white) (RZ)	50		0.346	0.359			
4000K（冷白色）（RL） 4000K (cool white) (RL)	40		0.380	0.380			
3500K（白色）（RB） 3500K (white) (RB)	35		0.409	0.394			
3000K（暖白色）（RN） 3000K (warm white) (RN)	30		0.440	0.403			
2700K（白炽灯色）（RD） 2700K (incandescent lamp) (RD)	27		0.463	0.420			
	P27		0.458	0.410			

特殊显色指数R9>0（饱和红色）

Special color rendering index R9 > 0 (deep red)

灯的初始光效 Lamp's initial efficiency		
等级 Grade	光效（lm/W） Efficiency (lm/W)	
	色调代码：65,50，40 Tone code: 65, 50, 40	35, 30, 27
Ⅰ	100	95
Ⅱ	85	80
Ⅲ	70	65

灯的光通维持率 Lamp's lumen maintenance rating		
宣称平均寿命（h） Nominal average lifetime (h)	3000h的光通维持率（%） Lumen maintenance rating at 3000h (%)	6000h的光通维持率（%） Lumen maintenance rating at 6000h (%)
25000	95.8	91.8
30000	96.5	93.1
35000	97.0	94.1
40000	97.4	94.8
45000	97.7	95.4
50000	97.9	95.8

注：

1.灯的平均寿命：不应低于25000h。

Notes:

1. A lamp's average lifetime should not be less than 25000h.

2.灯的功率因数（λ）：$p \leqslant 5W$的$\lambda \geqslant 0.4$；$p>5W$的$\lambda \geqslant 0.7$；宣称高功率因数的$\lambda \geqslant 0.9$。

3.灯的谐波电流：应符合GB17625-1-2012的要求。

4.型号示例：BPZ500-S840.E27 光通量为500lm，半配光型，*Ra*=80, 4000K，E27灯头。

2. A lamp's nominal power factor (λ): when $p \leqslant 5W$, $\lambda \geqslant 0.4$; when $p > 5W$, $\lambda \geqslant 0.7$; for lamp with high power factor, $\lambda \geqslant 0.9$.

3. A lamp's harmonic current should conform to stipulations of GB17625-1-2012.

4. Example: BPZ500-S840.E27, luminous flux 500lm, semi spatial light, *Ra*=80, 4000K, Lamp cap E27.

GB/T 31112-2014《普通照明用非定向自镇流LED灯规格分类》

序号 SN	名称 Name	注解 Annotation
1	定向照明 directional lighting	灯在立体角为π sr（相当于120°锥角圆锥的立体角）内有80%以上的光输出； Light output of the lamp in the solid angle of π sr (equals solid angle of 120° taper angle cone) is more than 80%
2	非定向照明 non-direct lighting	灯在立体角为π sr内的光输出小于80% Light output of the lamp in the solid angle of π sr is less than 80%
3	灯的配光分类：全配光型 Ominidirect light distribution （代码O）	灯的光束角大于180°，且光分布在0°~135°区域内任一角度的光强与该区域平均光强偏差不超过20%； Beam angle of the lamp is larger than 180°, and deviation between intensity of light distributed within the 0°~135° range from any angle and the average light intensity within the range is no larger than 20%
4	准全配光型 Quasi Ominidirectional（代码Q）	灯的光束角大于180°，光分布不符合上述要求。 Beam angle of the lamp is larger than 180° and light distribution doesn't meet the above requirements
5	半配光型 Semi spatial light（代码S）	灯的光束角不大于180°的非定向配光型。 Nondirectional light distribution with beam angle of the lamp no larger than 180°

1.5 LED照明产业现状

Current Situations of LED Lighting Industry

1.5.1 蓝光LED获2014年诺贝尔物理学奖：

2014年10月7日，瑞典皇家科学院宣布，将2014年诺贝尔物理学奖联合授予日本科学家的赤崎勇(Isamu Akasaki)、天野浩(Hiroshi Amano)以及美国加州大学圣巴巴拉分校的美籍日裔科学家中村修二(Shuji Nakamura)，表彰他们发明一种新型高效节能光源[蓝色发光二极管(LED)]，为节省能源开拓了新空间。

首次实现了氮化镓的PN结，为利用氮化镓材料制造蓝色发光二极管奠定了基础。发明的蓝光LED技术为研发明亮而节能的灯具，更高效的照明技术铺平了道路。将新开发的蓝光LED光源与已有的红光与绿光LED光源结合，人们终于可以通过三原色原理产生更加自然和实用的白光照明光源。

1.5.1 Blue LED won 2014 Nobel Prize in Physics:

On October 7, 2014, Royal Swedish Academy of Sciences announced that 2014 Nobel Prize in Physics was awarded to Japanese scientists Isamu Akasaki and Hiroshi Amano and Shuji Nakamura, Japanese-American professor at University of California, Santa Barbara for their invention of blue light-emitting diodes (LED), an efficient and energy-saving light source that expands new space for energy efficiency.

The discovery of GaN p-n junction made it possible to produce blue LED by using GaN materials. The blue LED technology paved the way to bright and energy-saving lamps and to more efficient lighting technologies. Putting together the blue LED that had been discovered lately and red and green LEDs that has been discovered before, people can produce more natural and practical white lighting in RGB color model.

1.5.2 中国照明学会主办的“中照照明奖”

2008~2014年，中国照明学会主办评选了“中照照明奖室内外工程设计奖”，其中，24个LED照明设计项目被评为室内外工程设计奖一等奖或特等奖[具体项目详见附件：2008~2014年中照照明奖——工程设计奖（24个LED照明一等奖）]。这个奖项是国家科技奖励办正式批准的中国照明领域唯一科技奖项。

1.5.2 CIES Lighting Awards

From 2008 to 2014 China Illumination Engineering Society sponsored "CIES Lighting Awards——Indoor/Outdoor Engineering Design Awards". 24 LED lighting design projects were elected first prizes or special prizes of Indoor/Outdoor Engineering Design Awards (See the attachments for detailed information: Attachments: 2008-2014 CIES Lighting Awards——Engineering Design Awards (24 LED lighting first prizes). It is the only technology award in the lighting field in China that are formally approved by National Office for Science & Technology Awards.

附件：2008~2014年中照照明奖——工程设计奖　Attachment: 2008~2014 CIES Lighting Awards——Engineering Design Awards

序号 SN	获奖项目 Projects	获奖单位 Awarded Enterprises	年份 Year	奖项 Awards	等级 Grade	图片 Photos
1	无锡五印坛城室内照明工程 Wuyintan City Interior Lighting Engineering in Wuxi	锐高照明电子（上海）有限公司 TridonicAtco Lighting Electronics (Shanghai)Co. Ltd.	2013年	工程设计奖（室内） Engineering Design Award (Interior)	一等奖 First Prize	
		上海艾特照明设计有限公司 Shanghai Aite Lighting Design Co. Ltd.				
		乐雷光电（上海）有限公司 Roled Optoele ctronics (Shanghai) Co.Ltd.				

（续表）（Continued）

序号 SN	获奖项目 Projects	获奖单位 Awarded Enterprises	年份 Year	奖项 Awards	等级 Grade	图片 Photos
2	上海星联科研大厦1号楼室内亮化工程 Interior Lighting Engineering of Building No.1 in Shanghai Star Alliance Science Research Mansion	飞利浦（中国）投资有限公司 Philips (China) Investment Co. Ltd.	2013年	工程设计奖（室内） Engineering Design Award (Interior)	一等奖 First Prize	

（续表）（Continued）

序号 SN	获奖项目 Projects	获奖单位 Awarded Enterprises	年份 Year	奖项 Awards	等级 Grade	图片 Photos
3	人民大会堂万人大礼堂照明光源节能改造工程 Light Source Energy-saving Renovation Project of the 10000-seat Grand Auditorium of the Great Hall of the People	北京星光影视设备科技股份有限公司 Beijing Xingguang Film&TV Equipment Technologies Co. Ltd. 东莞勤上光电股份有限公司 Dongguan Kingsun Optoelectronic Corp. 河北立德电子有限公司 Hebei Lide Electronics Co.Ltd. 浙江北光照明科技有限公司 SKYLED Lighting Technology Zhejiang Co., Ltd. 人民大会堂管理局 Administration of the Great Hall of the People 中国建筑科学研究院等 China Academy of Building Research etc.	2013年	工程设计奖（室内） Engineering Design Award (Interior)	一等奖 First Prize	

（续表）（Continued）

序号 SN	获奖项目 Projects	获奖单位 Awarded Enterprises	年份 Year	奖项 Awards	等级 Grade	图片 Photos
4	上海虹桥机场西航站楼及附属楼室内照明工程 Interior Lighting Project of West Terminal and Ancillary Building of Shanghai Hongqiao Airport	华东建筑设计研究院有限公司 East China Architectural Design & Research Institute Co. Ltd. 北京富润成照明系统工程有限公司 Beijing Fortune Lighting System Engineering Co. Ltd.	2010年	工程设计奖 Engineering Design Award	一等奖 First Prize	
5	上海世博轴夜景照明工程 Shanghai Expo Axis Nightscape Lighting Project	上海世博土地控股有限公司 Shanghai Expo Land Holdings Co.Ltd. 上海现代设计集团华东建筑设计研究院有限公司 East China Architectural Design & Research Institute Co.Ltd. of Shanghai Xian Dai Architectural Design Group 上海市政工程设计研究总院 Shanghai Municipal Engineering Design Institute (Group) Co. Ltd. 上海广茂达光艺科技股份有限公司 Shanghai Grandar Opto Technology Corp.	2010年	工程设计奖 Engineering Design Award	一等奖 First Prize	

（续表）（Continued）

序号 SN	获奖项目 Projects	获奖单位 Awarded Enterprises	年份 Year	奖项 Awards	等级 Grade	图片 Photos
6	无锡灵山二期梵宫室内照明工程 Interior Lighting Project of Buddhist Palace (Phase II) in Lingshan, Wuxi	上海复舜信息科技有限公司（艾特照明工作室） Shanghai Fushun Information Technology Co. Ltd. (Aite Lighting Studio)	2009年	工程设计奖 Engineering Design Award	一等奖 First Prize	
		华东建筑设计研究院有限公司 East China Architectural Design & Research Institute Co.Ltd.				
		锐高照明电子（上海）有限公司 TridonicAtco Lighting Electronics (Shanghai)Co. Ltd.				
7	同济大学教学科研综合楼“异形体空间”室内照明工程 Irregular Shape Space Interior Lighting Project in the Teaching and Research Complex of Tongji University	飞利浦（中国）投资有限公司 Philips (China) Investment Co. Ltd.	2008年	工程设计奖 Engineering Design Award	一等奖 First Prize	

（续表）（Continued）

序号 SN	获奖项目 Projects	获奖单位 Awarded Enterprises	年份 Year	奖项 Awards	等级 Grade	图片 Photos
8	苏州科技文化艺术中心夜景照明工程 Nightscape Lighting Project of Suzhou Sci-tech Culture & Art Center	上海莱亭景观工程有限公司 Shanghai Laiting Landscape Engineering Co.Ltd.	2008年	工程设计奖 Engineering Design Award	一等奖 First Prize	
		飞利浦（中国）投资有限公司 Philips (China) Investment Co.Ltd.				
9	南昌市一江两岸夜景照明工程 Nanchang Riverside Nightscape Lighting Project	广州凯图电气股份有限公司 Guangzhou Keatoo Electric Corp.	2014年	工程设计奖（室外） Engineering Design Award (Outdoor)	一等奖 First Prize	
		北京新时空照明技术有限公司 Beijing New Space New Lighting Technology Co. Ltd.				
		北京清华同衡规划设计研究院有限公司THUPDI				
		上海光联照明科技有限公司 Shanghai Grand Light Lighting Technology Co. Ltd.				

（续表）（Continued）

序号 SN	获奖项目 Projects	获奖单位 Awarded Enterprises	年份 Year	奖项 Awards	等级 Grade	图片 Photos
10	成都水井坊C区夜景照明工程 Chengdu Swellfun District C Nightscape Lighting Project	中辰照明 Aurora Enterprise	2014年	工程设计奖（室外）Engineering Design Award (Outdoor)	一等奖 First Prize	
11	天津团泊新城--萨马兰奇纪念馆夜景照明工程 Samaranch Memorial Museum Nightscape Lighting Project in Tuanbo New City，Tianjin	上海碧甫照明工程设计有限公司 Brandston Partnership Inc.（BPI）（Shanghai） 光缘（天津）科技发展有限公司 Guangyuan (Tianjin) Technology Development Co. Ltd.	2014年	工程设计奖（室外）Engineering Design Award (Outdoor)	一等奖 First Prize	

（续表）（Continued）

序号 SN	获奖项目 Projects	获奖单位 Awarded Enterprises	年份 Year	奖项 Awards	等级 Grade	图片 Photos
12	北京兴创大厦夜景照明工程 Beijing Xingchuang Tower Nightscape Lighting Project	豪尔赛照明技术集团有限公司 Haoersai Lighting Technology Group Co. Ltd.	2014年	工程设计奖（室外） Engineering Design Award (Outdoor)	一等奖 First Prize	
13	常州东经120景观塔夜景工程 Changzhou Dongjing 120 View Tower Nightscape Lighting Project	常州市城市照明工程有限公司 Changzhou Urban Lighting Engineering Co. Ltd. 北京清华同衡规划设计研究院有限公司 THUPDI	2014年	工程设计奖（室外） Engineering Design Award (Outdoor)	一等奖 First Prize	

（续表）（Continued）

序号 SN	获奖项目 Projects	获奖单位 Awarded Enterprises	年份 Year	奖项 Awards	等级 Grade	图片 Photos
14	北京营城建都滨水绿道夜景照明工程 Beijing Urban Construction —Riverside Waterfront Greenways Nightscape Lighting Project	深圳市高力特实业有限公司 Shenzhen GGE Lighting Co. Ltd.	2014年	工程设计奖（室外） Engineering Design Award (Outdoor)	一等奖 First Prize	
15	郑州会展宾馆 Zhengzhou Conference & Exhibition Hotel	河南新中飞照明电子有限公司 Henan Xinzhongfei Lighting Electronics Co. Ltd.	2013年	工程设计奖（室外） Engineering Design Award (Outdoor)	一等奖 First Prize	

（续表）（Continued）

序号 SN	获奖项目 Projects	获奖单位 Awarded Enterprises	年份 Year	奖项 Awards	等级 Grade	图片 Photos
16	台儿庄古城重建项目（二期亮化工程设计） Taierzhuang Ancient City Restoration Project (Phase 2 Lighting Engineering Design)	北京中辰泰禾照明电器有限公司——张帆工作室 Beijing Zhongchentaihe Lighting Electric Appliance Co.Ltd. (Zhang Fan Studio)	2013年	工程设计奖（室外） Engineering Design Award (Outdoor)	一等奖 First Prize	
17	湖州喜来登月亮酒店建筑景观照明工程 Huzhou Sheraton Moon Hotel Architectural Landscape Lighting Project	黎欧思照明（上海）有限公司 LEOX Design Partnership (Shanghai) 宁波华强灯饰照明有限公司 Ningbo Huaqiang Lamps& Lighting Co. Ltd. 上海光联照明科技有限公司 Shanghai Grand Light Lighting Technology Co. Ltd.	2013年	工程设计奖（室外） Engineering Design Award (Outdoor)	一等奖 First Prize	

（续表）（Continued）

序号 SN	获奖项目 Projects	获奖单位 Awarded Enterprises	年份 Year	奖项 Awards	等级 Grade	图片 Photos
18	大同市云冈石窟园区照明工程 Yungang Grottoes Park Lighting Engineering in Datong	北京豪尔赛照明技术有限公司 Beijing Haoersai Lighting Technology Co. Ltd. 雷士(北京)光电工程技术有限公司 NVC (Beijing) Optoelectronic Engineering Technology Co. Ltd.	2013年	工程设计奖（室外）Engineering Design Award (Outdoor)	一等奖 First Prize	
19	西安楼观台道教文化区夜景照明工程 Xi’an Louguan Tai Taoism Culture Park Nightscape Lighting Project	北京广灯迪赛照明设备安装工程有限公司 Beijing Guangdeng Disai Lighting Equipment Installation Engineering Co.Ltd.	2012年	工程设计奖（室外）Engineering Design Award (Outdoor)	一等奖 First Prize	

（续表）（Continued）

序号 SN	获奖项目 Projects	获奖单位 Awarded Enterprises	年份 Year	奖项 Awards	等级 Grade	图片 Photos
20	重庆园博园夜景照明工程 Chongqing Garden Expo Nightscape Lighting Project	北京良业照明技术有限公司 Beijing Landsky Lighting Engineering Co. Ltd.	2012年	工程设计奖（室外） Engineering Design Award (Outdoor)	一等奖 First Prize	
21	西安“天人长安塔”夜景照明工程 Xi’an Tianren Changan Tower Nightscape Lighting Project	碧谱照明设计（上海）有限公司 Brandston Partnership Lighting design Inc. (Shanghai)	2012年	工程设计奖（室外） Engineering Design Award (Outdoor)	一等奖 First Prize	

（续表）（Continued）

序号 SN	获奖项目 Projects	获奖单位 Awarded Enterprises	年份 Year	奖项 Awards	等级 Grade	图片 Photos
22	国家游泳中心夜景照明工程 National Aquatics Center Nightscape Lighting Project	北京市国有资产经营有限责任公司国家游泳中心 Beijing State-Owned Assets Management Co.Ltd. National Aquatics Center Branch	2012年	工程设计奖（室外） Engineering Design Award (Outdoor)	一等奖 First Prize	
23	北京宛平城地区夜景照明工程 Beijing Wanping City Area Nightscape Lighting Project	北京海兰齐力照明设备安装工程有限公司 Beijing Hailan Qili Lighting Equipment Installation Engineering Co.Ltd. 库柏电气（上海）有限公司 Cooper China Co. Ltd	2012年	工程设计奖（室外） Engineering Design Award (Outdoor)	一等奖 First Prize	

（续表）(Continued)

序号 SN	获奖项目 Projects	获奖单位 Awarded Enterprises	年份 Year	奖项 Awards	等级 Grade	图片 Photos
24	世博中国国家馆夜景照明工程 Shanghai Expo China Pavilion Nightscape Lighting Project	上海城市之光灯光设计有限公司 Shanghai Citelum Lighting Design Co.Ltd	2011年	工程设计奖（室外） Engineering Design Award (Outdoor)	一等奖 First Prize	
		欧司朗（中国）照明有限公司 OSRAM (China) Lighting Co.Ltd.				

1.5.3 LED产品竞争力和分布

1.LED灯比白炽灯节电90%，LED灯比节能灯节电30%。

2.LED灯售价每年下降20%左右；3~5年LED灯的价格会更低。

3.芯片材料成本占LED封装成本的40%，下降空间较小。

4.封装材料成本占LED成本的60%，下降空间较大。

5.上游：LED芯片和封装材料企业，主要在珠三角、长三角、环渤海等地区。

6.下游：LED灯具企业，主要在珠三角、长三角地区。

7.中国上万家照明企业，生产的LED产品产量占全球的60%。

1.5.3 Competitiveness and distribution of LED products

1. LED lamps use 90% less electric power than incandescent lamps, and 30% less power than compact fluorescent lamps.

2. LED lamp's price will drop about 20% every year. In 3-5 years, its price will be much lower.

3. Costs of chip are 40% of the packaging costs, and can be hardly dropped.

4. Costs of packaging material are 60% of LED costs, and can be much less.

5. Upstream: LED chip and packaging material companies. They are mainly located at Pear River Delta, Yangtze River Delta Bohai Sea Ring Area.

6. Downstream: LED lamp companies. They are mainly at Pearl River Delta and Yangtze River Delta.

7. There are tens of thousands lamp producers in China. They produce 60% LED lamps in the world.

LED产品竞争力成熟度三项指标
Three indexes indicating mature competitiveness of LEDs

指标 Index	标准 Standard
视觉指标 Visual index	显色指数超过80以上 *CRI* ≥ 80
性能指标 Performance index	光效超过170 lm/W Efficient > 170 lm/W

1.5.4 中国LED照明产业情况
1.5.4 Situations of LED lighting industry in China

LED照明产品替换率 LED lighting products' rate of substitution		
国家 Country	替换率 Rate of substitution	占有率 Market share
日本 Japan	80%	
美国 USA	30%	50%
中国 China	20%	
欧洲 Europe	60%	

1.5.5 广东省LED照明产业的情况
1.5.5 Situations of LED lighting industry in Guangdong Province

2014年第1~4季度广东LED产业专利申请分布图
Patent Application Distribution of LED Industry in Guangdong Q1~Q4, 2014

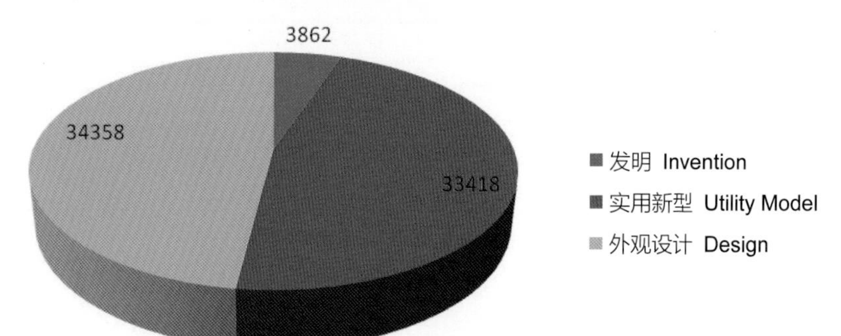

截止到2014年4季度累计专利授权类型 Total Patent Authorization Types up to Q4, 2014				
	总数 Total	发明 Invention	实用新型 Utility model	外观设计 Design
广东LED LED in Guangdong	71638	3862	33418	34358
全国LED LED in China	244924	22316	129032	93576
		0.05391	0.466484	0.479606

2014年第1~4季度广东LED重点领域出口构成图

Composition Diagram of Major LED Exports in Guangdong Q1~Q4, 2014

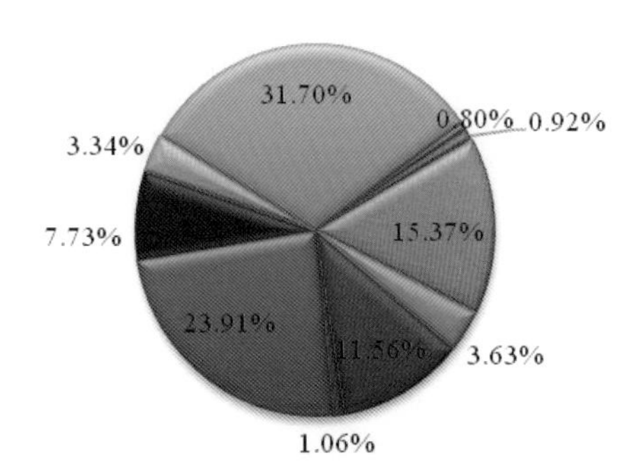

- LED其他彩色监视器 Other LED color monitors
- 发光二极管 LED
- LED枝形吊灯 LED chandeliers
- 圣诞树用的成套LED灯具 Lighting sets of a kind used for Christmas trees
- LED非电气灯具及照明装置 non-electric LED lamps and lighting fittings
- 装有发光二极管的显示板 Display panel with LED
- LED相关配件、材料 LED related parts and materials
- 电气的LED台灯、LED床头灯和LED落地灯 Electric LED desk, bedside and floor-standing lamps
- LED其他电灯及照明装置 Other LED lamps and lighting fittings
- LED发光标志、发光铭牌及类似品 LED illuminated signs, illuminated name-plates and the like

	出口额（亿元） Value of export (RMB100 million)	同比增长（%） Year-on-year growth rate (%)
LED其他彩色监视器 Other LED color monitors	144.33	30.58
装有发光二极管的显示板 Display panel with LED	34.08	18.47
发光二极管 LED	108.50	37.95
LED相关配件、材料 LED related parts and materials	9.91	-14.88
LED枝形吊灯 LED chandeliers	224.43	34.81
电气的LED台灯、LED床头灯和LED落地灯 Electric LED desk, bedside and floor-standing lamps	72.53	43.80

（续表）（Continued）

	出口额（亿元） Value of export (RMB100 million)	同比增长（%） Year-on-year growth rate (%)
圣诞树用的成套LED灯具 Lighting sets of a kind used for Christmas trees	31.31	-2.23
LED其他电灯及照明装置 Other LED lamps and lighting fittings	297.55	47.62
LED非电气灯具及照明装置 non-electric LED lamps and lighting fittings	7.47	21.48
LED发光标志、发光铭牌及类似品 LED illuminated signs, illuminated name-plates and the like	8.64	4.18

2014年第1~4季度广东省LED总产值构成表（单位：亿元）
Composition Table of LED Gross Output Value in Guangdong Province Q1~Q4, 2014 (Unit: RMB 100 million)

	指标值（亿元） Index value (RMB 100 million)	同比增长率（%） Year-on-year growth rate (%)	占总产值比例（%） Percentage to the gross output value of that period (%)
LED芯片外延片 LED chip wafers	14.34	23.68	0.41
LED封装元器件 Electronic components for LED packaging	512.93	21.10	14.82
LED背光源 LED back light	258.27	18.42	7.46
LED照明灯具 LED lamps	772.41	39.04	22.32
LED光源及专业灯具 LED light sources and professional lamps	427.90	28.58	12.37
LED灯饰 LED decorative lamps	214.91	6.87	6.21
LED显示屏 LED displays	287.61	30.07	8.31
LED配件、材料 LED parts and materials	625.25	7.55	18.07
LED装备 LED equipment	87.28	-2.49	2.52
LED生产性服务业 LED production services	259.16	46.97	7.49

广东省LED照明产业的情况

Situations of LED lighting industry in Guangdong Province

2014年第1~4季度广东省LED产业总产值构成图

Composition Diagram of LED Industry's Gross Output Value in Guangdong Province Q1~Q4, 2014

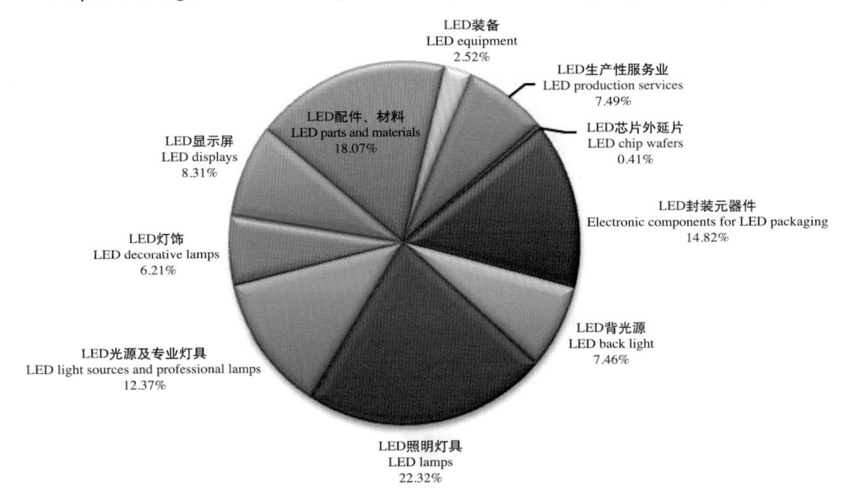

1.5.6 住房城乡建设部科研课题《中国建筑电气与智能化节能发展报告(2014)》

1.5.6 Scientific study of Ministry of Housing and Urban-Rural Development Report on the Development of Electric and Intelligent Energy-saving in China Buildings (2014)

目录

Contents

1.6 LED照明产业发展趋势

Development Trends of LED Lighting Industry

发展趋势之一：政策扶持

Trend I: Policies support

序号 SN	时间 Time	国家政策支持内容 Details of supportive national policies
1	2013年	六部委联合发布《半导体照明节能产业规划》，提出LED三大发展目标：节能减排效果及市场份额；产业规模及重点企业实力；技术创新能力及标准检测认证体系；在上述三方面稳步增长。 Planning of Semiconductor Lighting Energy Saving Industry jointly released by the six ministries and commissions set out three targets: Effect of energy saving and emission reduction & market share; industrial scale and major enterprises' capability; technical innovation capability & test, inspection and certification systems. The industry will grow in the above aspects.
2	2011年	科技部颁布《关于印发国家十二五科学和技术发展规划的通知》，节能环保位居七大战略性新兴产业之首，其中，LED照明又位居四大节能环保技术之首。 Ministry of Science and Technology issued Notification about Issuance of Planning of Scientific & Technical Development in the Twelfth Five-Year Plan. Energy-saving and environmental protection is on top of the seven strategical new industries. And, LED lighting is on top of the four energy-saving and environmental protection technologies.
3	2010年	国务院常务会议，审议并原则通过《国务院关于加快培育和发展战略性新兴产业的决定》，确定选择节能环保、新一代信息技术、生物、高端装备制造等七个战略新兴产业，LED 产业作为节能环保的重要产业会成为扶持对象。 The meeting of the standing committee of the State Council reviewed and generally approved The State Council's Decision about Speeding Up the Development of Strategical New Industries, and selected seven strategic new industries: energy saving and environmental protection, new generation information technology, biology, high level equipment manufacturing, etc. As an important industry of energy saving and environmental protection, LED industry will have the supports.

发展趋势之二：资金扶持
Trend 2: Financial support

序号 SN	时间 Time	国家资金扶持内容 Financial support
1	2013年	国家发改委出台“十二五”LED节能产业规划国家将对LED相关设备实行免税政策。 According to LED energy saving industry planning for the Twelfth Five-year-plan Period issued by National Development and Reform Commission, the government will have tax exemption policies for equipment related to LEDs.
2	2012年	财政部、工信部、海关总署、国税总局等四部委发布《关于调整重大技术装备进口税收政策有关目录的通知》。该通知中，新增了28项重大技术装备，海工装备、太阳能应用、LED设备等三大新兴产业成了受益最为明显的板块。 Ministry of Finance, Ministry of Industry and Information Technology, General Administration of Customs and State Administration of Taxation jointly issued Notification about Revised Catalog of Major Technology and Equipment Import Taxation Policy. In the Notification, 28 technologies and equipment are added. Marine engineering equipment, solar energy application and LED equipment the three new industries will obviously be benefited.
3	2011年	中国环资工委提出在“十二五”期间将安排80亿人民币中央财政预算外资金采购LED节能产品，同时将带动不低于30亿元人民币的地方配套资金。 China Environment Resource and Energy Saving System Innovation Work Committee announced to use RMB 8 billion extra budgetary fund to procure LED energy saving products in the Twelfth Five-year-plan Period, which would drive no less than RMB 3 billion local fund.

发展趋势之三：技术发展

Trend 3: Technological development

序号 SN	技术 Technology	描述 Description
1	新型衬底外延芯片技术 New lining, wafer and chip	采用碳化硅、氮化镓等新型衬底开发外延芯片技术进一步提高LED芯片发光效率及可靠性。 Use Sic, GaN and other new lining to develop wafter chip technology, and further increase LED chip efficiency and reliability.
2	倒装芯片新技术 Flip chip	采用倒装技术，降低热阻和封装成本。 Use flip chip technology to reduce thermal resistance and packaging costs.
3	晶元级封装新技术 Wafer-level packaging	缩小封装体积改善散热，提升整体器件可靠度。 Reduce package size and improve thermal passage, and make overall components more reliable.
4	宽色域、高显色指数的荧光粉新技术 Fluorescence with wide color gamut and high CRI	色彩还原性好，荧光转换效率高。 Better colorreproducivity and high fluorescence conversion efficiency.
5	高效LED驱动电源新技术 LED drive power source	通过降低磁性材料在高频情况下的磁损耗，开发更高效率的半导体驱动芯片，以提高转换效率。 By reducing magnetic material's wearing in high frequency, develop LED drive chip with higher efficiency so as to increase conversion rate.
6	LED智能驱动芯片 LED smart drive chip	为了适应物联网和智能照明技术的发展，智能驱动芯片有极大发展空间。 To catch up with the development of Internet of Things and intelligent lighting technology, smart drive chip has much room for development.

发展趋势之四：产品发展——替代传统照明
Trend 4: Product development —— Replacement of traditional lighting

序号 SN	国家 Country	LED替代传统照明 LED's replacement of traditional lighting
1	美国 USA	加州2012年1月1日起，发布禁令的适用范围将扩大至超过9寸长的高强度放电灯及节能灯，2014年1月1日起，将禁用范围进一步涵盖美国国家监管的普通照明白炽灯及增强光谱灯。 Since January 1, 2012, high intensity discharge lamps and CFLs more than 9 inches long would be banned in California. From January 1, 2014, common incandescent lamps and enhanced spectroscopic lamps that were under national supervision of USA would be banned.
2	澳大利亚 Australia	2007年2月，决定于2010年停止使用普通白炽灯，取而代之的是LED照明用品，澳大利亚成为世界上第一个计划禁止使用传统白炽灯的国家。 In February 2007 Australia decided to stop using common incandescent lamps in 2010, and use LED lighting as replacement. Australia is the first country that plans to ban traditional incandescent lamps.
3	加拿大 Canada	2007年4月25日，自然资源部部长加里·伦恩宣布，加拿大定于2012年开始禁止销售白炽灯，成为继澳大利亚后第二个宣布将禁用白炽灯的国家。 On April 25, 2007, Gary Lunn, Minister of Natural Resources, announced that Canada decided to ban the sales of incandescent lamps in 2012. Following Australia, Canada was the second country that would ban incandescent lamps.
4	欧盟 European Union	首脑会议于2008年达成协议，决定欧盟各国将逐步用节能灯取代白炽灯；欧盟各国也拟通过立法从2009年开始禁止生产白炽灯泡。欧盟成员国能源部长要求欧盟委员会在2008年底前制订计划，从2010年起禁止在欧盟销售包括白炽灯在内的高耗能家用照明设备。 European Council reached an agreement in 2008, and decided that the member countries would gradually replace incandescent lamps with CFLs. The member countries planned to ban the production of incandescent lamps in 2009 by means of legislation. Energy Ministers of the member countries asked European Council to draft plans before the end of 2008, and stop the sales of household lighting equipment with high energy consumption, including incandescent lamps, in European Union since 2010.

（续表）(Continued)

序号 SN	国家 Country	LED替代传统照明 LED's replacement of traditional lighting
5	阿根廷 Argentina	2008年签署法案，决定从2011年起彻底禁止普通灯泡的使用。 It signed an act in 2008 to prohibit the use of common lamps since 2011.
6	俄罗斯 Russia	2009年11月建议，从2010年1月开始，逐步禁止使用白炽灯，改用节能灯。俄罗斯将在2012年到2014年内彻底禁用25瓦至75瓦的白炽灯。目前，莫斯科已开始采取措施，让老式白炽灯退出市场。 It was proposed in November 2009 that starting from January 2010 incandescent lamps would be banned step by step and people would start to use CFLs. Russia would totally ban 25W~75W incandescent lamps from 2012 to 2014. At present, Moscow has taken measures to put old incandescent lamps out of the market.
7	新西兰 New Zealand	能源部2007年表示，将从2009年开始，禁止使用白炽灯泡。 The Ministry of Energy indicated in 2007 that incandescent lamps would be prohibited since 2009.

中国自2014年10月1日起，将禁止进口和销售60W及以上普通照明白炽灯（白炽灯淘汰路线图详见下表）。

Starting from October 1, 2014, China will not allow import and sales of 60W and above incandescent lamps for common lighting. (The chart below shows how will incandescent lamps be out of use)

步骤 Step	实施期限 Period	目标产品 Target products	额定功率 Rated power	实施方式 Implementation	备注 Remarks
1	2011.11.1- 2012.9.30	过渡期 Transitional period			发布公告及路线图 Issue public announcements and route maps
2	2012.10.1起 Since Oct. 1, 2012	普通照明白炽灯 Incandescent lamps for common lighting	≥100W	禁止进口、销售 Import and sales prohibited	——
3	2014.10.1起 Since Oct. 1, 2014	普通照明白炽灯 Incandescent lamps for common lighting	≥60W	禁止进口、销售 Import and sales prohibited	——
4	2015.10.1- 2016.9.30	进行中期评估，调整后续政策 Mid-term evaluation for policy adjustment			

（续表）(Continued)

步骤 Step	实施期限 Period	目标产品 Target products	额定功率 Rated power	实施方式 Implementation	备注 Remarks
5	2016.10.1起 Since Oct. 1, 2016	普通照明白炽灯 Incandescent lamps for common lighting	≥15W	禁止进口、销售 Import and sales prohibited	最终禁止的目标产品和时间，以及是否禁止生产视中期评估结果而定。In the end what products will be prohibited, when will these product be prohibited and whether the products will be allowed for production or not will be subject to the result of mid-term evaluation.

发展趋势之五：市场发展——LED照明市场前景

Trend 5: Market development —— Outlook of LED lighting market

产业联盟的白皮书的结论：目前选定LED照明色温的光源质量不差于传统的节能照明光源的质量。

Conclusion in the white book of the industrial league: The quality of lighting of selected LED color temperature is not worse that of than the traditional CFLs.

LED照明在建筑室内和室外的应用：中国室内照明产品目前以改造工程视觉效果替换光源为主，目的是节能，但有些场所不注意LED照明的光色应用，如：色温、显色指数的选择不当，舒适度受影响。中国户外景观亮化工程已全面采用LED照明，能够实现艺术设计效果，有安全、可塑性强、视觉感受逼真、节能环保、价格比较低等好处。

LED lighting's indoor and outdoor applications: Indoor lamps in China are mainly substituting lamps to improve visual effects and save energy. Some places have not paid sufficient attentions to the color of LED lighting, for example, color temperature and CRI are not properly selected, so that people may not feel very comfortable. For outdoor landscape lighting in China, LED lighting is widely used. It can show the effect of art designs, give true visual sense, and is safe, flexible, energy saving, environmental friendly and less expensive.

1.全球照明用电量约占19%，利用LED灯可节约40%能耗，每年减少二氧化碳5.55亿吨。中国照明用电量约占12%。

The global power consumption of lighting is around 19%. If LED lamps are used, we can save about 40% energy, and cut down 555 million tons of carbon dioxide every year. In China lighting uses about 12% of the total power.

2. 2013年，中国LED产业整体规模约2576亿元，同比增长率约34%。

In 2013 the total output of LED industry in China was about RMB 257.6 billion, at a year-on-year growth rate of about 34%.

3.2013年LED灯市场占有率约8.9%，封装器件价格平均降幅达20%，接近市场承受的临界点。

In 2013 LED lamps took about 8.9% of the market. The prices of packaging components dropped 20% in average, close to the acceptable scope of the market.

4.2013年LED照明备案立项投资总额约208亿元，比去年增长15.9%。

In 2013 the registered investment to LED lighting projects were about RMB 20.8 billion in total, at an increase of 15.9% than the year before.

5.预测2014年中国LED产业增长率约40%。

It was forecasted that LED industry in China will increase about 40% in 2014.

6.预测2015年中国LED照明市场占有率约20%。

It was forecasted that LED lamps will have about 20% of the market share in 2015.

7. 预测2020年中国照明的80%将被LED照明所代替。

It was forecasted that 80% of traditional lamps in China will be replaced with LED lamps in 2020.

1.7 未来LED照明创新技术
LED Lighting Innovation Technologies in Future

LED创新技术之一：解决飞机时差的LED照明创新技术；

Innovation Technology 1: Solve Jet Lag

LED创新技术之二：解决集中精力的LED照明创新技术；

Innovation Technology 2: Promote Concentration

LED创新技术之三：解决睡觉失眠的LED照明新创技术；

Innovation Technology 3: Alleviate Insomnia

LED创新技术之四：解决果蔬保鲜的LED照明创新技术；

Innovation Technology 4: Fruits and Vegetables Fresh-Keeping

LED创新技术之五：促进蔬菜生长的LED照明创新技术；

Innovation Technology 5: Promote Vegetables Growth

LED创新技术之六：解决交通站牌的LED照明创新技术；

Innovation Technology 6: Transport Station Boards

LED创新技术之七：解决产品形态的LED照明创新技术；

Innovation Technology 7: Diversify Product Forms

LED创新技术之八：解决健康照明（舒适）的LED照明创新技术；

Innovation Technology 8: Healthy Lighting (Comfort)

LED创新技术之九：解决智能控制（系统）的LED照明创新技术；

Innovation Technology 9: Intelligent Control (System)

LED创新技术之十：解决可见光通信LED照明创新技术。

Innovation Technology 10: Visible Light Communication

LED创新技术之一：解决飞机时差的LED照明创新技术

Innovation Technology 1: Solve Jet Lag

通过LED光源的色温（在2700K~6500K之间选择最合适色温)、颜色（选择最合适的五种颜色，例如，日出采用柔和的蓝色和淡紫色模拟天空；早餐采用暖色光源模拟餐厅烛光；白天采用灯光全亮）、控制（多种情景灯光控制模式：叫醒、早餐、白天、正餐、睡觉）、音乐（适于睡眠轻音乐）等四个方面解决飞机的时差问题。

Recover from jet lag by means of LED lighting's color temperatures (choose the most suitable color temperature at 2700K~6500K), colors (choose the five most suitable colors, for example, use soft blue and light purple to simulate the sky at sunrise, use warm lighting to simulate restaurant candle light at breakfast, and use bright light for the day), controls (optional lighting control modes for different scenarios: wake-up call, breakfast, daylight, meals and sleep) and music (light music suitable for sleep).

LED创新技术之二：解决集中精力的LED照明创新技术

Innovation Technology 2: Promote Concentration

通过LED光源的色温（在2700K~6500K之间选择最合适的三种色温）、颜色（选择适合学习的颜色）、控制（三种情景灯光控制模式：准备、学习、休息）等三个方面解决学习集中的问题。

Concentrate your mind by means of LED lighting's color temperatures (choose the three most suitable color temperatures at 2700K~6500K), colors (choose the color(s) suitable for study) and controls (optional lighting control modes for three scenarios: preparation, study and break).

LED创新技术之三：解决睡觉失眠的LED照明新创技术

通过LED光源的色温（在2700K~6500K之间选择最合适的两种色温）、颜色（选择最合适的颜色）、控制（二种情景灯光控制模式：准备、睡觉）、音乐（适合睡觉的音乐）等四个方面解决失眠或神经衰弱的问题。另外，房间一面墙的LED大屏幕的可根据需求变化（春、夏、秋、冬）。

Innovation Technology 3: Alleviate Insomnia

Treat insomnia or panasthenia by means of LED lighting's color temperatures (choose two color temperatures at 2700K~6500K), colors (choose the most suitable color), controls (optional lighting control modes for two scenarios: preparation and sleep) and music (music suitable for sleep). Moreover, LED screen on one wall can change as needed (spring, summer, autumn and winter).

LED创新技术之四：解决果蔬保鲜的LED照明创新技术

Innovation Technology 4: Fruits Fresh—Keeping

通过LED光源的波长（通过实验得到适合保鲜的波长）、色温和显色指数（在2700K~6500K之间选择最合适的色温、显色指数，使果蔬更加美观）、控制（不同果蔬采用不同的灯光控制模式）等三个方面解决果蔬保鲜且美观的问题。

Keep fruits and vegetables fresh and look good by means of LED lighting's wavelengths (wavelengths suitable for keeping fresh can be obtained through lab tests), color temperatures and CRIs (choose the most suitable color temperatures at 2700K~6500K and CRI to let fruits and vegetables look good) and controls (apply different lighting control modes for different fruits and vegetables).

LED创新技术之五：促进植物生长的LED照明创新技术

Innovation Technology 5: Promote Plants Growth

通过LED光源的波长（通过实验得到适合植物快速增长的波长）、控制（不同植物采用不同的灯光控制模式）等两个方面解决植物快速增长且美观的问题。目前，实验结果比植物正常增长快2/3。

Let plants grow fast and look good by means of LED lighting's wavelengths (wavelengths for plants' fast growth can be obtained through lab test) and controls (apply different lighting control modes for different plants). In current experiments, plants with LED lighting grow 2/3 faster than those without LED lighting.

LED创新技术之六：解决交通站牌的LED照明创新技术

Innovation Technology 6: Transport Station Boards

通过LED光源的颜色（选择组合最合适的颜色，使站牌更加直观，通过GPS显示下一辆公车的具体位置）、控制（不同车辆采用不同的灯光控制模式）等两个方面解决公共交通站牌和车辆时间间隔的问题。

Show public transport station boards and intervals between buses on the bus stop sign by means of LED lighting's color (choose the most suitable colors so that the station boards can be read easily. Show location of the next bus with data from GPS) and controls (apply different lighting control modes for different buses).

LED创新技术之七：解决产品形态的LED照明创新技术

LED照明与建筑形式的结合不理想，多数LED照明均沿用传统概念的照明光源形状；由于LED照明具有体积小、重量轻、易控制等特点，可代替传统照明灯具的形式，形态上有很大的发展空间，在产品形态上可延伸开发、创新。

Innovation Technology 7: Diversify Product Forms

LED lightings are not properly designed for buildings. For buildings, as most LED lightings work with shapes in traditional concepts. As LED lightings are small, light and easy for control, they can replace traditional lamps to offer diversified developments. The forms of lighting can be further developed and innovated.

LED创新技术之八：解决健康照明（舒适）的LED照明创新技术

Innovation Technology 8: Healthy Lighting (Comfort)

可开发LED功能性照明技术，通过改变光谱等参数模拟自然光环境，以满足人的舒适性要求。

LED functional lighting technologies can be developed and make people more comfortable by simulating natural lighting with modified spectrum.

LED创新技术之九：解决智能控制（系统）的LED照明创新技术

Innovation Technology 9: Intelligent Control (System)

发挥LED容易实现调节光的亮度，调节光的颜色，调节光的色温等特性，研发性价比高的LED智能控制系统，实现情景照明模式。

Fully utilize LED's features of adjustable brightness, color and color temperature to invent LED smart control system with high performance-price ratio and offer different lighting modes for different scenarios.

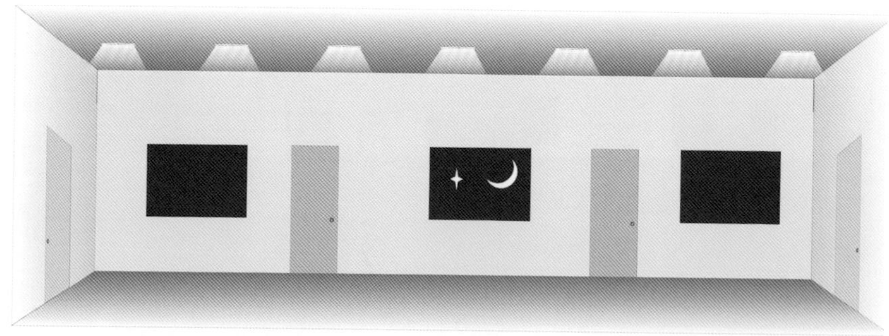

Light brightness keep 10%~50% brightness in the standby mode.
无人时自动调节成微亮状态

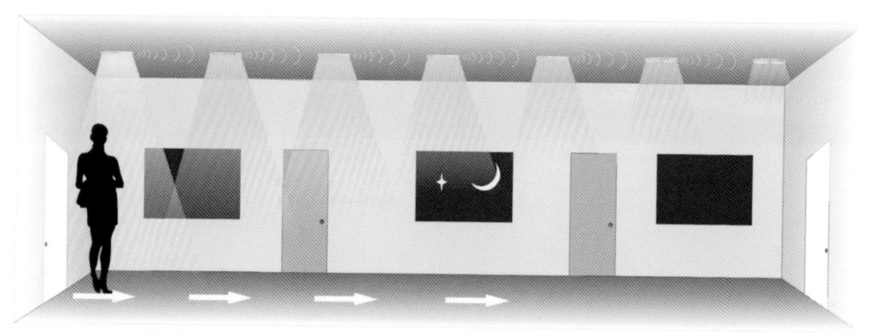

Light up to 100% brightness when a moving person is detected.
有人时智能点亮

LED创新技术之十：解决可见光通信LED照明创新技术

Innovation Technology 10: Visible Light Communication

利用现有的有线基础设施，将数据发送给LED灯，并作为无处不在的新接入点，LED灯同时实现照明和通信功能。预测未来光通信市场规模约60亿美金。

Send data through existing wired infrastructure to LED lamps, and use them as new AP. The LED lamps can be used for both lighting and communications. It is forecasted that the optical communication market will worth USD 6 billion in the future.

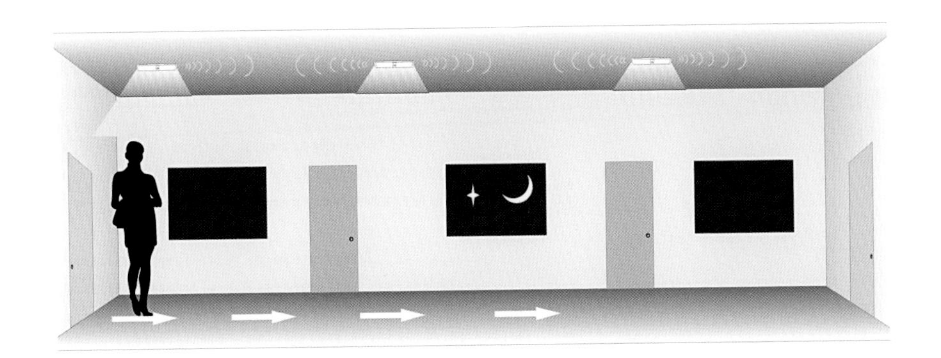

当首灯亮起时，通过2.4G无线通信控制后面灯具陆续点亮。

When the first lamp lights up, the next lamps will light up one after another through 2.4G wireless control.

参考文献:

1.欧阳东《BIM技术——第二次建筑设计革命》（中国建筑工业出版社）2013.7

2.中国照明学会官方网站“中国照明网”

3.广东省半导体照明产业联合创新中心“2013年广东省LED产业运行监测报告”。

4.《普通照明用非定向自镇流LED灯性能要求》GB/T 24908-2014

5.《普通照明用非定向自镇流LED灯规格分类》GB/T 31112-2014

6.任元会《中国建筑电气协会第三届建筑电气及智能化节能技术发展论坛品牌颁奖会及技术交流会》（2014年北京）上发言

References:

1.Ouyang Dong. BIM Technology——The Second Revolution in Architectural Design. China Architecture & Building Press. July 2013.

2. www.lightingchina.com. The official website of China Illuminating Engineering Society.

3. "2013 Guangdong Province LED Industry Monitoring Report" by Guangdong Solid State Lighting Industry Innovation Center.

4. Non-directional Self-ballasted LED-lamps for General Lighting Services—Performance Requirements GB/T 24908-2014.

5. Classification of Non-directional Self-ballasted LED-lamps for General Lighting Services GB/T 31112-2014.

6. Speech by Ren Yuanhui at "Awarding and Symposium of the 3rd Building Electric & Intelligent Energy Saving Technology Development Forum of China Building Electrics Society" (2014, Beijing)

LED照明技术

LED Lighting Technology

2.1 LED照明技术特点

Technical Features of LED Lighting

技术特点之一：衬底技术

Feature 1: Substrate Technique

序号 SN	名称 Name	技术特点 Technical features	图形 Photo
1	蓝宝石衬底 Sapphire substrate	蓝宝石衬底的成分是三氧化二铝（Al_2O_3）；氮化镓（GaN）基器件的外延层在高温下生长在蓝宝石衬底上。蓝宝石具有生产技术成熟、化学稳定性好、易于清洗等优点，但存在晶格失配较大、热导性差等缺点。 Sapphire substrate’s component is aluminum oxide (Al_2O_3). GaN based device’ wafer is grown on sapphire substrate in high temperature. Sapphire features in mature production technology, stable chemical characteristics and easy washing. But, it has poor lattice match and weak thermal conductivity.	
2	碳化硅衬底 SiC substrate	碳化硅衬底的成分是碳化硅（SiC）单晶体，氮化镓（GaN）基器件的外延层在高温下生长在碳化硅衬底上。采用碳化硅衬底制备的LED芯片是垂直结构芯片，碳化硅衬底具有晶格匹配度好、热导性好等优势，但其价格昂贵。 SIC substrate’s component is crystalline silicon carbide. GaN based device's wafer is grown on SiC substrate. LED chip made from SiC substrate is a vertical structure. SiC substrate features in good lattice match and good thermal conductivity. But it is expensive.	

（续表）（Continued）

序号 SN	名称 Name	技术特点 Technical features	图形 Photo
3	同质衬底 Native substrate	同质衬底的成分是氮化镓（GaN）单晶体，氮化镓（GaN）基器件的外延层在高温下生长在氮化镓衬底上。GaN衬底采用HVPE(氢化物气相外延)、氨热法或钠流法生长。同质GaN衬底具有晶格匹配、导热性好、外延技术简单等优势，缺点是制备困难，价格昂贵。 Native substrate's component is crystalline GaN. GaN based device's wafer is grown on GaN substrate in high temperature. GaN substrate is grown in HVPE method, ammonothermal method or Na-flux method. Native substrate GaN wafer features in good lattice match, good thermal conductivity and simple water technology. But, it is difficult to produce and its price is high.	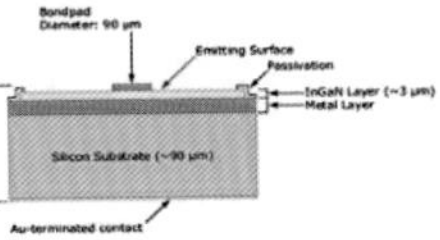
4	硅衬底 Si substrate	硅衬底的成分是硅（Si）单晶体，氮化镓（GaN）基器件的外延层在高温下生长在硅衬底上。硅衬底具有衬底材料便宜的优势，缺点是晶格失配、热稳定性差。 Silicon substrate's components is crystalline silicon. GaN based device's wafer is grown on silicon substrate in high temperature. Silicon wafter features in cheap substrate material. It has poor lattice match and poor thermal stability.	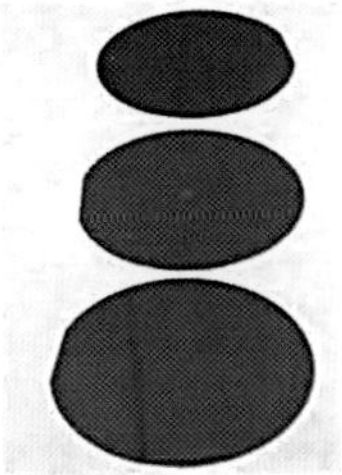

技术特点之二：外延技术

Feature 2: Wafer Technique

序号 SN	名称 Name	技术特点 Technical features	图形 Photo
1	量子阱 Quantum well	量子阱是由两种不同的半导体材料相间排列形成的、具有明显量子限制效应的电子或空穴的势阱。特点是材料厚度小于电子的德布罗意波长。氮化镓LED的辐射复合发生在量子阱。 Quantum well, formed by putting two different semi-conductive materials next to each other, is electrons or holes with distinctive quantum confinement effect. The materials thickness is less than de Brogile wavelength. GaN LED's radiative recombination occurs at quantum well.	
2	P型掺杂 P-type doping	P型掺杂是在氮化镓材料中掺入少量其他元素或化合物（一般掺镁），其主要载流子为空穴。通过退火活化等技术改善P型掺杂，提高载流子浓度和迁移率。 P-type doping is to add small amount of other element or composition (usually magnesium) in GaN material. Its main carrier is hole. Annealing, activation and other approaches are adopted to improve P-type doping so that carrier's concentration and transfer rate will be increased.	

（续表）(Continued)

序号 SN	名称 Name	技术特点 Technical features	图形 Photo
3	缓冲层 Buffer	缓冲层（在蓝宝石衬底上生长）是在异质外延中的插入层，可以起到改善晶格失配、提高晶体质量、减小应力等效果。 Buffer (grown on sapphire substrate) is a layer in non-native wafer. It can improve lattice mismatch and crystalline quality, and reduce stress.	P-GaN外延层 量子阱 GaN缓冲层 蓝宝石基板
4	图形化衬底 Pattern Sapphire Surface	图形化衬底（PSS—Pattern Sapphire Surface）是通过在蓝宝石衬底表面制作细微结构的图形，然后在这种图形化的衬底表面进行LED材料外延，提升LED发光亮度。 Patter Sapphire Surface (PSS) is to make micro patterns on the surface of sapphire substrate, and grow LED wafer on the pattern surface so as to increase LEDs brightness.	

技术特点之三：芯片技术

Feature 3: Chip Technique

序号 SN	名称 Name	技术特点 Technical features	图形 Photo
1	正装结构 Upright structure	正装结构（又称为水平芯片）是LED的两个电极在芯片的同侧，电流水平流动。优点是工艺成熟、成本低，缺点是大电流下发光效率衰减快。 In upright structure (also called horizontal chip) the two electrodes of LED are at the same side, and current flows horizontally. The advantages are mature process and low costs. The disadvantage is that lighting efficiency declines quickly in large current.	
2	倒装结构 Inverted structure	倒装结构LED芯片的出光面在蓝宝石面，可有效提升器件的散热能力和出光效率、改善出光光型。优点是光型好、出光效率高、散热好，缺点是工艺复杂、成本高。 In inverted structure, the LED chip's lighting comes from the side of sapphire, which can increase the components' heat dissipation and lighting efficiency, and can improve the light shape. The advantages are better light shape, higher lighting efficiency, and better heat dissipation. The disadvantages are complicated process and high cost.	
3	垂直结构 Vertical structure	垂直结构是LED的两个电极在芯片的上下两侧，电流垂直流动。优点是导热性好，适合做大功率芯片，缺点是工艺复杂、成本高。随着设备和工艺不断成熟，转移衬底和碳化硅衬底越来越接近量产。 In vertical structure, the LED's two electrodes are at top and bottom sides of the chip so that the current moves vertically. The advantage lies in better heat conduction, which makes the structure suitable for large power chip. The disadvantages are complicated process and high costs. While the equipment and process become more and more mature, transferred substrates and SiC substrates are getting closer to mass production.	

技术特点之四：封装技术
Feature 4: Packaging Technique

序号 SN	名称 Name	技术特点 Technical features	图形 Photo
1	直插式封装 Dual Inline-pin Package（DIP）	直插式封装（DIP—Dual Inline-pin Package）是两根引线、封装材料为环氧树脂，将小功率芯片固定在支架中的封装体。特点：应用于小功率照明（<0.2W）和信号指示等。优点：价格低，缺点：散热差、应用范围窄。始于20世纪60年代、最早的封装技术。Dual Inline-pin Package (DIP) is a package to fix low power chip on support by using two lead wires and epoxy resin packaging materials. Features: It is applied in low power lighting (<0.2W) and signaling. Advantage: Low price. Disadvantage: Poor thermal dissipation, limited applications. It was invented in the 1960s, and is the earliest packaging technology.	
2	表贴式封装 Surface Mounted Device (SMD)	表贴式封装（SMD—Surface Mounted Device)是直接将芯片贴在带热沉的水平支架上，封装材料为硅胶，将不同功率的芯片固定在支架中的封装体。特点：散热好，方便量产。优点：价格低、体积小、光效高、寿命长。缺点：光形不好，不太适合大电流(>350mA)驱动。成熟于20世纪90年代的封装技术。Surface Mounted Device (SMD) is a package to put chips directly on horizontal supports with silica gel packaging material. Features: Good head dissipation and easy mass production. Advantages: Low price, small size, high efficiency and long life. Disadvantages: Bad lighting shape, and not suitable for large current (>350mA) driving. The packaging technology became mature in the 1990s.	

（续表）(Continued)

序号 SN	名称 Name	技术特点 Technical features	图形 Photo
3	集成式封装 Chip On Board (COB)	集成式封装（COB—chip on board）是将多颗中小功率芯片直接封装在金属基印刷电路板（MCPCB—Metal Core Printed Circuit Board）上的封装体。优点：减少支架制造工艺及成本，减少热阻。缺点：标准化较困难。成熟于2005年后的封装技术。 Chip On Board (COB) is a package in which several middle or small power chips are directly put on Metal Core Printed Circuit Board (MCPCB). Advantages: Reduce the processes and costs of support, and reduce thermal resistance. Disadvantage: It is difficult to make it a standard. It became mature after 2005.	
4	倒装式封装 Flip Chip Package (FCP)	倒装式封装（FCP—Package）是将倒装芯片直接焊接在陶瓷或硅基板上的封装体。优点：体积小、光效高、光形好。缺点：封装设备贵、专利壁垒较高、芯片来源不稳定。成熟于2013年后的封装技术。 Flip Chip Package (FCP) is a package in which flip chip is directly welded on ceramics or silicon substrates. Advantage: Small size, high efficiency and good lighting shape. Disadvantage: Expensive packaging equipment, high patent barrier and unstable source of chips. It became mature after 2013.	

技术特点之五：散热器技术

Feature 5: Radiator Technique

序号 SN	名称 Name	内容 Content	技术特点 Technical features	图形 Photo
1	金属散热 Heat dissipation through metal	铝合金 Aluminum alloy	铝合金散热器（导热系数约237W/m²·k）是把光源产生的热量通过辐射和传导到空气中。优点：加工性能好、美观，缺点：不绝缘。 Aluminum alloy radiator (thermal conductivity efficiency about 237W/m²·k) radiates and conducts the heat produced by light source to the air. Advantages: Good processing performance and beautiful form. Disadvantage: Not insulated.	
		铁合金 Ferrous alloy	铁合金散热器（导热系数约80W/m²·k）是把光源产生的热量通过辐射和传导到空气中。优点：成本较低，缺点：加工性差、导热性差、不绝缘 Ferrous alloy radiator (thermal conductivity efficiency about 80W/m²·k) radiates and conducts the heat produced by light source to the air. Advantage: Low costs. Disadvantage: Poor processing, poor heat conductivity and not insulated.	

（续表）(Continued)

序号 SN	名称 Name	内容 Content	技术特点 Technical features	图形 Photo
2	非金属散热 Heat dissipation through non-metal	复合陶瓷 Ceramic matrix composite	复合陶瓷散热器（导热系数约175W/m^2·k）把光源产生的热量通过辐射和传导到外界。优点：绝缘性好，缺点：加工较困难，成本高。 CMC radiator (thermal conductivity about 175W/m^2·k) radiates and conducts the heat produced by light source to the environment. Advantage: Good insulation. Disadvantage: Difficult processing and high costs.	
		复合塑料 Plastic composite	复合塑料散热器（导热系数约100W/m^2·k）把光源产生的热量通过辐射和传导到外界。优点：成本低，易加工，缺点：导热性差，易老化。 Plastic composite radiator (thermal conductivity about 100W/m^2·k) radiates and conducts the heat produced by light source to the environment. Advantage: Low costs and easy processing. Disadvantage: Poor thermal conductivity and easy aging.	
		玻璃 Glass	玻璃散热器（导热系数约41W/m^2·k）把光源产生的热量通过辐射和传导到外界。优点：成本低、易加工，缺点：易碎、导热性差。 Glass radiator (thermal conductivity about 41W/m^2·k) radiates and conducts the heat producted by light source to the environment. Advantage: Low costs and easy processing. Disadvantage: Fragile and poor thermal conductivity.	

技术特点之六：驱动技术
Feature 6: Drive Technique

序号 SN	分类 Type	名称 Name	技术特点 Technical features	图形 Photo
1	直流驱动 DC drive	恒流模式 Current Control	恒定电流（简称CC）——输出二作，输出恒定电流给LED以保持稳态工作。优点：发光稳定，LED利用率高，单管功率恒定。缺点：电路复杂，成本较高。适用范围：广泛。 Current Control (CC in short) output mode gives constant current to LED to keep it in stable state. Advantage: Stable lighting, high LED utilization and constant power for single diode. Disadvantages: Complicated circuit and high costs. Applicable scope: Wide.	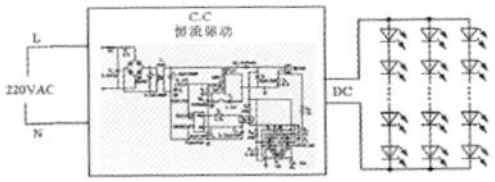
2		恒压模式 Voltage Control	恒定电压（VC）——输出工作，输出恒定电压通过电阻器限流后提供给LED工作。优点：发光稳定，可使用标准化电源，单个电源可驱动多个灯具。缺点：供电半径短。适用范围：灯条、灯带等小功率灯具。 Voltage Control (VC) output mode gives constant voltage that passes resistor for current limitation and offers current for LED. Advantages: The lighting is stable, standard power source can be used, and one power source can drive several lamps. Disadvantages: The power supply radius is limited. Application scope: light bars strips and other small power lamps.	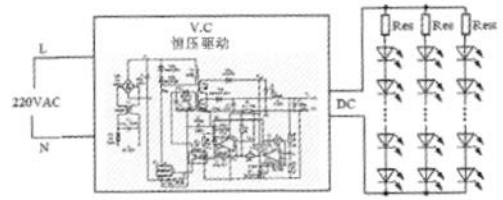
3		恒流+恒压式 Current Control + Voltage Control	恒流恒压（CC+VC）输出工作，用恒压源作为集中供电电源，同时用灯具上的恒流源解决出光稳定性问题。优点：大功率电源转换效率高、电源稳定性好，单个电源可驱动多个灯具、驱动成本低。缺点：适用范围窄。适用范围：多个同类灯具照明。即提供恒定输出功率。 In CC+VC output mode, constant voltage power source is used as central power source, and a constant current power source on the lamp will guarantee stable lighting. Advantages: Conversion rate for large-power power source is high, power source is quite stable, one power source can drive several lamps so that the drive costs are low. Disadvantage: Limited application scope. Application scope: Lighting for several lamps of the same type. That is, it can offer constant output power.	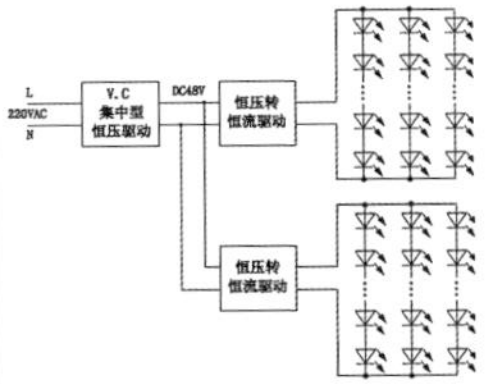

（续表）（Continued）

序号 SN	分类 Type	名称 Name	技术特点 Technical features	图形 Photo
4	交流驱动 AC drive	直接驱动 Direct drive	通过多组LED串联，实现交流直接驱动。 It can directly drive several groups of LEDs in series connection with AC current. 优点：免驱动接入市电，价格低。缺点：频闪，受电压波动影响大，LED芯片利用率较低。适用范围：小功率紧凑型光源。 Advantages: LED lamps can directly connect to municipal power grid without drive, and the costs are low. Disadvantages: There are strobes, it is largely affected by the fluctuation of voltage, and LED chip's utilization rate is low. Application scope: Low-power compact light source.	220VAC PTC LED1 LED2 LED94
5		分段驱动 Multi-stage drive	根据正弦波形不同区段，分别驱动不同串联数量的LED芯片。优点：芯片利用率较高，抗电网波动能力较强，缺点：频闪。适用范围：小功率紧凑型光源。 Use different stages of sine waveform to separately drive LED chips that are series connected. Advantages: Chip's utilization is high, and it is very resistant to the fluctuation of grid. Disadvantage: Strobe. Application scope: Low-power compact light source.	EMI 元件

技术特点之七：配光技术

Feature 7: Light Distribution Technique

序号 SN	名称 Name	技术特点 Technical features	图形 Photo
1	透射配光 Distribution through transmission	通过光学透镜实现光源配光。优点：光场形状易控制，缺点：面光源配光困难，材料受限。适用范围：广泛。 Light is distributed through lens. Advantage: It is difficult to control the shape of light field. Disadvantage: It is difficult to distribute area light source, and the materials are limited. Application scope: Wide.	
2	反射配光 Distribution through reflection	通过反射面实现光源配光。优点：成本低，适用范围广。缺点：光场形状不易控制，体积大。适用范围：广泛。 Light is distributed through reflector. Advantages: Low costs and wide application scope: Disadvantages: It is difficult to control the shape of light field, and the size of reflector is large. Application scope: Wide.	
3	面板配光 Distribution through light guide plate	通过导光板实现光源配光，优点：眩光少，较宽配光。缺点：光损较大。适用范围：较窄。 Light is distributed through light guide plate. Advantages: Less glare, and wide distribution. Disadvantage: Lots of light loss. Application scope: Limited.	
4	组合配光 Distribution through combined lens and reflector	通过光学透镜和反射面实现光源配光。优点：易实现个性化配光，缺点：配光复杂，适用范围：较窄。 Light is distributed through lens and reflector. Advantage: Customized light distribution can be easily realized. Disadvantage: The distribution is complicated. Application scope: Quite limited.	

2.2 LED照明节能特性

Energy Saving Features of LED Lighting

LED节能照明的定义：

用高亮度发光二极管作为发光源，光效高、低能耗、寿命长、易控制、低运维、安全环保、光谱丰富、绿色环保，是固态冷光源。

Definition of LED energy-saving lighting:

Use bright light-emitting diode as the light source. It features high efficiency, low energy consumption, long lifetime, easy control, low maintenance costs, and rich spectrum. It is safe and environment friendly. It is a solid-state cold light source.

序号 SN	特性 Features	内容 Content
综合节能特性之一 Energy Saving Feature 1	光效高 High Luminous Efficiency	电光转换效率约30%，发光效率约100lm/W Energy conversion efficiency is about 30%, and luminous efficacy is about 10lm/W.
综合节能特性之二 Energy Saving Feature 2	能耗低 Low Energy Consumption	耗电量低，例如：单只9W LED灯（相当于60W白炽灯，15W节能灯）每天8小时年耗电 Low power consumption. For example, one 9W LED (the luminous efficacy of which equals to that of a 60W incandescent lamp or a 15W CFL) working 8 hours per day uses about 26KWh electricity in a year.
综合节能特性之三 Energy Saving Feature 3	寿命长 Long Life	使用寿命约50000小时，是白炽灯使用寿命的50倍，荧光灯使用寿命的10倍。 Its lifetime is about 50,000 hours, which is 50 times of that of incandescent lamp or 10 times of fluorescent lamp.
综合节能特性之四 Energy Saving Feature 4	易控制 Easy Control	通过BIM技术和智能控制技术，实现智慧照明。从而起到二次节能和按需照明的目的。 Through BIM technology and intelligent control technology, we can achieve intelligent lighting and reach the goals of further energy saving and lighting by the demand.
综合节能特性之五 Energy Saving Feature 5	低运维 Easy Operation & Maintenance	易实现自动巡检，更换频率低，节约人工成本。 It can do automated regular self-checking so that the replacement will be less often, and labor costs can be reduced.
综合节能特性之六 Energy Saving Feature 6	安全环保 Safe and Clean	直流驱动无频闪，光源不含紫外线，发热低不易产生安全隐患。 There is no strobe in DC driving mode. The light source does not have ultraviolet radiation. It produces less heat, and causes less potential safety hazards.
综合节能特性之七 Energy Saving Feature 7	光谱丰富 Broad Spectum	易实现色彩变化，适用于城市亮化工程。更适用于特定光谱场合，如农业照明，紫外固化等。 Change of colors can be easily done, which make it suitable for lighting projects in cities. It is more suitable for places that require specific spectrum, like agricultural lighting and UV curing.
综合节能特性之八 Energy Saving Feature 8	绿色环保 Green & Environment Friendly	不含汞，易回收。 It is free of Mercury and can be easily recycled.

LED照明技术标准现状与发展规划

Current Situations and Development Plan of LED Lighting Technical Standards

3.1 国外LED照明技术标准现状

Current Situations of Foreign LED Lighting Technical Standards

欧洲的LED相关标准

1.2009年09月，欧盟开始执行EN62471:2008标准。

2.2013年9月，欧盟颁布《LED照明产品最新能效要求》。

3.IEC 62504：一般照明，发光二极管（LED）产品及相关设备，术语和定义。

Europe standards related to LED

1. In September 2009, EU started the practice of EN 62471:2008.

2. In September 2013, EU issued The Latest Energy-Efficiency Requirements for LED Lighting Products.

3.IEC 62504: General lighting – Light emitting diode (LED) products and related equipment – Terms and definitions (draft).

Edition 1.0 2009-08

TECHNICAL REPORT

colour inside

Photobiological safety of lamps and lamp systems –
Part 2: Guidance on manufacturing requirements relating to non-laser optical radiation safety

INTERNATIONAL
ELECTROTECHNICAL
COMMISSION

PRICE CODE X

ICS 29.140

美国的LED相关标准

U.S. standards related to LED
IES RP-16, Definitions
ANSI BSR C78.377A,Chromaticity
IES LM 79, Luminous Flux
IES LM 80, Lumen Depreciation
IES TM-21, LED Lifetime
IES LM 82, Characterization of LED Light Engines and LED Lamps as a Function of Temperature

加拿大的LED相关标准

2012年12月，由加拿大自然资源学会制定了《关于LED灯具的技术指南》标准，并开始实施。

Canadian standards related to LED

In December 2012, Natural Resources Canada drafted and implemented Model Technical Specifications for Procurement of LED Luminaires in Canada - A Guide.

Model Technical Specifications for Procurement of LED Luminaires in Canada—A Guide
Version 1.0

LightSavers Canada
The Canadian Urban Institute

December 31, 2012

Generously supported by:

Natural Resources Canada　Ressources naturelles Canada

日本的LED相关标准

2004年,由日本照明学(JIES)、日本照明委员会(JCIE)、日本照明器具工业会(JIL)以及日本电球工业会(JEL)制订出四团体共同标准《照明用白色LED测光方法通则》(JIS C 8152-2007)。

Japanese standards related to LED

In 2004 JIES, JCIE, JIL and JEL jointly drafted Measuring Methods of White Light Emitting Diode for General Lighting (JIS C 8152-2007).

日本工業規格 JIS C 8152 : 2007

照明用白色発光ダイオード（LED）の測光方法

Measuring methods of white light emitting diode for general lighting

1　適用範囲

この規格は，照明用途の白色の発光ダイオード（以下，LED という。）単体の CIE 平均化 LED 光度，全光束などの測光量及び全光束に対する光色に関する量を求める方法について規定する。

なお，この規格は，複数の照明用白色 LED を基板などに実装したもの（モジュール）についても適用できる。

2　引用規格

次に掲げる規格は，この規格に引用されることによって，この規格の規定の一部を構成する。これらの引用規格は，その最新版（追補を含む。）を適用する。

JIS C 1609-1　照度計　第 1 部：一般計量器

JIS Z 8113　照明用語

JIS Z 8724　色の測定方法－光源色

JIS Z 8725　光源の分布温度及び色温度・相関色温度の測定方法

JIS Z 8726　光源の演色性評価方法

JIS Z 8782　CIE 測色標準観測者の等色関数

CIE 127　Measurement of LEDs

3　用語及び定義

この規格で用いる主な用語及び定義は，**JIS Z 8113** によるほか，次による。

3.1

発光ダイオード，LED

電子流によって励起されたとき，光放射を放出する p-n 接合をもつ固体デバイス。LED 単体ともいう。

3.2

LED モジュール

LED 単体を基板などに実装するか，又は複数の LED を平面的若しくは立体的に配列して，機械的，電気的，及び光学的に多数の要素で構成して，一つのユニットとして取り扱えるようにしたもの，又はその集合体。

注記　LED モジュールは，LED 単体をアレイ状又はクラスタ状に配列したものも含む。

3.2 中国 LED照明技术标准现状

Current Situations of LED Lighting Technical Standards in China

2013年11月29日，国家颁布实施《建筑照明设计标准》GB50034-2013，2014年6月1日开始实施。

On November 29, 2013, the government published Standard for Lighting Design of Buildings (GB 50034-2013), which was implemented on June 1, 2014.

UDC

中华人民共和国国家标准

P

GB 50034-2013

建筑照明设计标准

Standard for lighting design of buildings

2013-11-29 发布

2014-06-01 实施

国家2014年9月16日颁布，且2015年8月1日实施的《普通照明用非定向自镇流LED灯性能要求》GB/T24908-2014）。

On September 16th, 2014, the government published Non-directional Self-ballasted LED-lamps for General Lighting Services—Performance Requirements (GB/T 24908-2014) which is to be implemented on August 1st, 2015.

ICS 29.140.99
K 71

中华人民共和国国家标准

GB/T 24908—2014
代替 GB/T 24908—2010

普通照明用非定向自镇流LED灯
性能要求

Non-directional self-ballasted LED-lamps for general lighting services
—Performance requirements

(IEC 62612:2009, Self-ballasted LED-lamps for general lighting services
—Performance requirements, NEQ)

2014-09-03 发布 2015-08-01 实施

中华人民共和国国家质量监督检验检疫总局
中国国家标准化管理委员会 发布

2008年底，国家颁布GB7000系列灯具国家标准（强制性）。

At the end of 2008, the government published GB 7000 national standards (compulsory).

GB 7000.1与IEC60598-1的对应关系

IEC60598-1 版本	IEC60598-1 发布年份	对应的国家标准
第7版	2008	目前尚无
第6版	2003	GB7000.1-2007，等同IEC
第5版	1999	GB7000.1-2002，等同IEC
第4版	1996	无
第3版	1992	GB7000.1-1996，等同IEC
第2版	1984	GB7000-86，参照IEC
第1版	1979	GB7000-86，参照IEC

ICS 29.140.01
K 72

GB

中 华 人 民 共 和 国 国 家 标 准

GB7000.1-2007

代替 GB7000.1-2002

灯具一般安全要求与试验

General safety requirements and tests for luminaires

(IEC60598-1:2006, luminaires - Part 1, General requirements and tests, IDT)

3.3 中国LED照明技术相关政策

Policies related to LED Lighting Technologies in China

2009年9月，国家发展和改革委员会联合科学技术部、工业和信息化部、财政部、住房和城乡建设部、国家质量监督检验检疫总局联合发布《半导体照明节能产业发展意见》。

In September 2009, National Development and Reform Commission, Ministry of Science and Technology, Ministry of Industry and Information Technology, Ministry of Finance Ministry of Housing and Urban-Rural Development, and General Administration of Quality Supervision, Inspection and Quarantine the six ministries jointly issued Opinions on the Development of LED Energy Saving Industry.

国家发展和改革委员会
科学技术部
工业和信息化部
财政部
住房和城乡建设部
国家质量监督检验检疫总局
文件

发改环资[2009]2441号

关于印发半导体照明节能产业发展意见的通知

各省、自治区、直辖市及计划单列市、副省级省会城市、新疆生产建设兵团发展改革委、经贸委（经委、经信委、工信委、工信厅）、科技厅（科委）、财政厅（局）、住房城乡建设厅（建委、建设局）、质量技术监督局：

为推动我国半导体照明节能产业健康有序发展，培育新的经济增长点，扩大消费需求，促进节能减排，国家发展改革委、科技部、工业和信息化部、财政部、住房城乡建设部、国家质检总局联合制定了《半导体照明节能产业发展意见》。现印发给你们，请结合实际贯彻落实。

附：半导体照明节能产业发展意见

2009年10月，国家发展改革委、联合国开发计划署、全球环境基金共同启动“中国逐步淘汰白炽灯、加快推广节能灯（PILESLAMP）”计划 。

In October 2009, National Development and Reform Commission, United Nations Development Programme and Global Environment Facility jointly started Phasing-out Incandescent Lamps & Energy Saving Lamps Promotion (PILESLAMP).

2010年4月，发展改革委、财政部、人民银行、税务总局四部门发布《关于加快推行合同能源管理促进节能服务产业发展的意见》。

In April 2010, National Development and Reform Commission, Ministry of Finance, The People's Bank of China and State Administration of Taxation published Opinions on Speeding up the Implementation of Contracted Energy Management to Promote the Development of Energy Saving Service Industry.

国务院办公厅文件

国办发[2010]25 号

国务院办公厅转发发展改革委等部门关于加快推行合同能源管理促进节能服务产业发展意见的通知

各省、自治区、直辖市人民政府，国务院各部委、各直属机构：

发展改革委、财政部、人民银行、税务总局《关于加快推行合同能源管理促进节能服务产业发展的意见》已经国务院同意，现转发给你们，请认真贯彻执行。

国务院办公厅

二○一○年四月二日

2010年10月，《国务院关于加快培育和发展战略性新兴产业的决定》将半导体照明列入我国战略性新兴产业：节能环保和新材料产业的重要发展方向。

In October 2010, Decision on Accelerating the Fostering and Development of New Strategic Industries by the State Council listed LED lighting as new strategic industry in China: It is the important development direction of energy saving, environment friendly and new material industry.

国务院关于加快培育和发展战略性新兴产业的决定

国发〔2010〕32号

各省、自治区、直辖市人民政府，国务院各部委、各直属机构：

战略性新兴产业是引导未来经济社会发展的重要力量。发展战略性新兴产业已成为世界主要国家抢占新一轮经济和科技发展制高点的重大战略。我国正处在全面建设小康社会的关键时期，必须按照科学发展观的要求，抓住机遇，明确方向，突出重点，加快培育和发展战略性新兴产业。现作出如下决定：

一、抓住机遇，加快培育和发展战略性新兴产业

战略性新兴产业是以重大技术突破和重大发展需求为基础，对经济社会全局和长远发展具有重大引领带动作用，知识技术密集、物质资源消耗少、成长潜力大、综合效益好的产业。加快培育和发展战略性新兴产业对推进我国现代化建设具有重要战略意义。

（一）加快培育和发展战略性新兴产业是全面建设小康社会、实现可持续发展的必然选择。我国人口众多、人均资源少、生态环境脆弱，又处在工业化、城镇化快速发展时期，面临改善民生的艰巨任务和资源环境的巨大压力。要全面建设小康社会、实现可持续发展，必须大力发展战略性新兴产业，加快形成新的经济增长点，创造更多的就业岗位，更好地满足人民群众日益增长的物质文化需求，促进资源节约型和环境友好型社会建设。

（二）加快培育和发展战略性新兴产业是推进产业结构升级、加快经济发展方式转变的重大举措。战略性新兴产业以创新为主要驱动力，辐射带动力强，加快培育和发展战略性新兴产业，有利于加快经济发展方式转变，有利于提升产业层次、推动传统产业升级、高起点建设现代产业体系，体现了调整优化产业结构的根本要求。

（三）加快培育和发展战略性新兴产业是构建国际竞争新优势、掌握发展主动权的迫切需要。当前，全球经济竞争格局正在发生深刻变革，科技发展正孕育着新的革命性

2010年11月，发展改革委、住房和城乡建设部、交通运输部三部委联合组织“半导体照明产品应用示范工程项目”，在全国共选择50个半导体照明应用项目开展示范。

In November 2010, National Development and Reform Commission Ministry of Housing and Urban—Rural Development and Ministry of Transport jointly organized "LED Lighting Product Application Demonstration Projects", and chose 50 LED lighting application project throughout the country as demonstration models.

国家发展和改革委员会办公厅
住房和城乡建设部办公厅 文件
交通运输部办公厅

发改办环资[2010]2082号

关于组织申报半导体照明产品应用示范工程项目的通知

各省、自治区、直辖市及计划单列市、副省级省会城市、新疆生产建设兵团发展改革委（经贸委、经信委）、住房城乡建设厅（建委、建设交通委、建设局、市政市容委）、交通运输厅（局、委）：

为贯彻落实《关于进一步加大工作力度确保实现“十一五”节能减排目标的通知》（国发〔2010〕12号）和国家发展改革委等六部委《关于印发半导体照明节能产业发展意见的通知》（发改环资〔2009〕2441号）要求，进一步推动绿色照明工程，促进半导体照明节能产业健康有序发展，国家发展改革委、住房城乡建设部、交通运输部决定组织开展半导体照明产品应用示范工程。现将有关事项通知如下：

一、在不同气候条件的地区，选择20个半导体室内照明应用项目、15个半导体路灯应用项目和15个半导体隧道灯应用项目开展示范。

二、每个省、自治区、直辖市及计划单列市、新疆生产建设兵团可根据《半导体照明产品应用示范工程实施方案》（见附件一）和《半导体照明产品技术要求（2010版）》（见附件三），针对每种类型产品应用示范可申报2个备选项目。

三、填写申报材料（见附件二）及其电子文档（光盘）共六份，并于9月30日前分别报送至国家发展改革委、住房城乡建设部和交通运输部各两份。

联系人及电话：

蒋炳尧　国家发展改革委环资司　（010）68505570
吴　博　　　　　　　　　　　　（010）68519036
严盛虎　住房城乡建设部城建司　（010）58933058
吕　娟　交通运输部政策法规司　（010）65293760

国家发展改革委办公厅
住房城乡建设部办公厅
交通运输部办公厅
二〇一〇年八月三十日

中国LED照明相关行业政策：2011年11月，国家发布《中国淘汰白炽灯路线图》。对作为白炽灯的替代产品之一高效节能LED照明产品具有重要意义，对培育半导体照明产品市场起到积极作用，也为我国半导体照明企业提供了巨大的发展空间。

Policies related to LED lighting industry in China: In November 2011, the central government published Route Map of Phasing-out Incandescent Lamps in China. It meant a lot to efficient, energy saving LED lighting, one of the replacements for incandescent lamp. It was positive for fostering the market of LED lighting products, and offered large room of the development of LED lighting companies in China.

3.4 LED照明技术发展规划

Development Plan of LED Lighting Technologies

1.国际上从事照明LED标准化研究的标准组织有国际电工委员会、国际照明委员会和各国对应的标准化组织及相关企业。国际电工委员会（IEC）和国际照明委员会（CIE）都非常关注LED 的发展及相关LED器件的标准化工作。

1. International organizations engaging in the study of LED lighting standards are International Electrotechnical Commission (IEC), International Commission on Illumination (CIE), and related standardization organizations and companies in the world. IEC and CIE are very concerned with LED's development and standardization of LED parts.

2.CIE曾经发表过LED检测方法的技术报告，由于近年来LED产品的技术发展迅速，CIE目前正在对测试方法标准进行修订。

2. CIE published technical reports about LED test methods. As LED product technologies develop quickly in recent years, CIE is revising the standards of test methods.

3.IEC近两年也加大了对LED标准的研究，相继对LED模块、LED连接件及LED控制件提出了标准草案。由于目前还没有统一的照明LED产品性能方面的国际标准，且各国LED的研究发展速度不同因此发达国家都在积极准备建立自己的LED标准体系。

3. IEC has done more studies into LED standards, putting forward draft standards for LED modules, LED fittings and LED controllers. As there is no uniform international standards about the performances of LED lighting products at present, and the countries in the world research and develop LEDs at different paces, developed countries actively prepare to build LED standard systems of their own.

4.美国正在根据照明LED的特性开展照明LED的技术标准和测试方法的研究。

4. The USA is making researches on LED lighting's technical standards and test methods according to LED lighting's features.

5.日本则将研究重点放在照明用白光LED的测试方法和技术标准上。

5. Japan is mainly making researches on test methods and technical standards for white LED lighting.

参考文献

1. 欧盟EN62471:2008标准

2. 欧盟《LED照明产品最新能效要求》

3. IEC 62504：一般照明，发光二极管（LED）产品及相关设备，术语和定义

4. IES RP-16, Definitions

5. ANSI BSR C78.377A, Chromaticity

6. IES LM 79, Luminous Flux

7. IES LM 80, Lumen Depreciation

8. IES TM-21, LED Lifetime

9. IES LM 82, Characterization of LED Light Engines and LED Lamps as a Function of Temperature

10. 加拿大自然资源学会《关于LED灯具的技术指南》标准

11. 日本照明学(JIES)、日本照明委员会(JCIE)、日本照明器具工业会(JIL)和日本电球工业会（JEL）四团体共同标准《照明用白色LED测光方法通则》（JIS C 8152-2007)

12. 建筑照明设计标准GB50034-2013

13. 普通照明用非定向自镇流LED灯性能要求GB/T24908-2014

14. GB 7000系列灯具国家标准（强制性）

15. 国家发展改革委联合科技部、工信部、住房和城乡建设部、财政部、质检总局《半导体照明节能产业发展意见》

16. 国家发展改革委、联合国开发计划署、全球环境基金“中国逐步淘汰白炽灯、加快推广节能灯计划”（PILESLAMP）

References

1. EU—EN 62471:2008 standard

2. EU The Latest Energy-Efficiency Requirements for LED Lighting Products

3. IEC 62504: General lighting – Light emitting diode (LED) products and related equipment – Terms and definitions (draft).

4. IES RP-16, Definitions

5. ANSI BSR C78.377A,Chromaticity

6. IES LM 79, Luminous Flux

7. IES LM 80, Lumen Depreciation

8. IES TM-21, LED Lifetime

9. IES LM 82, Characterization of LED Light Engines and LED Lamps as a Function of Temperature

10. Natural Resources Canada—Model Technical Specifications for Procurement of LED Luminaires in Canada - A Guide

11. JIES, JCIE, JIL and JEL—Measuring Methods of White Light Emitting Diode for General Lighting (JIS C 8152-2007)

12. Standard for Lighting Design of Buildings (GB 50034-2013)

13. Non-directional Self-ballasted LED-lamps for General Lighting Services—Performance Requirements (GB/T 24908-2014)

14. GB 7000 national standards (compulsory)

15. National Development and Reform Commission, Ministry of Science and Technology, Ministry of Industry and Information Technology, Ministry of Housing and Urban-Rural Development, Ministry of Finance and General Administration of Quality Supervision, Inspection and Quarantine—Opinions on the Development of LED Energy Saving Industry

16. National Development and Reform Commission, United Nations Development Programme and Global Environment Facility—Phasing-out Incandescent Lamps & Energy Saving Lamps Promotion (PILESLAMP)

17. National Development and Reform Commission, Ministry of Finance, The People's Bank of China and State Administration of Taxation—Opinions on Speeding up the Implementation of Contracted Energy Management to Promote the Development of

17. 发展改革委、财政部、人民银行、税务总局《关于加快推行合同能源管理促进节能服务产业发展的意见》

18.《国务院关于加快培育和发展战略性新兴产业的决定》（国发[2010]32号）

19. 发展改革委等三部“半导体照明产品应用示范工程项目”

20.《中国淘汰白炽灯路线图》

Energy Saving Service Industry

18. Decision on Accelerating the Fostering and Development of New Strategic Industries by the State Council (No.32 [2010] of the State Council)

19. National Development and Reform Commission and two other ministries—“LED Lighting Product Application Demonstration Projects”

20. Route Map of Phasing-out Incandescent Lamps in China

传统照明和LED照明的对比

Traditional Lighting VS LED Lighting

对比之一：光源性质关系

Difference I: Properties of Photo sources

	白炽灯 Incandescent lamp	荧光灯 Fluorescent lamp	LED LED lamp
图形比较 Photo			
特点比较 Features	1.光源特点：热辐射 Feature of light source: Heat radiation 2.光电转换率：低，发热量大 Photoelectric conversion efficiency: Low with lots of heat 3.光生物安全：无，有紫外线 Photobiological safety: Unsafe, UV available 4.环境污染：无 Environmental pollution: N/A	1.光源特点：气体放电 Feature of light source: Gas discharge 2.光电转换率：中 Photoelectric conversion efficiency: Medium 3.光生物安全：有紫外线，频闪严重 Photobiological safety: UV available, serious strobe 4.环境污染：汞污染（节能灯） Environmental pollution: Mercury contamination (energy-saving lamp)	1.光源特点：半导体固态 Feature of light source: Semiconductor solid-state 2.光电转换率：高 Photoelectric conversion efficiency: High 3.光生物安全：根据需求设置紫外线 Photobiological safety: Offer UV as circumstances require 4.环境污染：无 Environmental pollution: N/A
对比结果 Conclusion	LED安全，无污染，光电转换效率高，采用LED是必然的。 Due to the safety, pollution-free and high photoelectric conversion efficiency of LED, it is inevitable to employ LED.		

对比之二：节能降耗
Difference 2: Energy–Saving and Consumption Reduction

	白炽灯 Incandescent lamp	荧光灯 Fluorescent lamp	LED LED lamp
图形比较 Photo			
特点比较 Features	1.光效:低（2%~3%） Luminous efficiency: Low (2%~3%) 2.平均寿命:1000小时。 Average life: 1,000 hours. 3.运维费:高 Operation and maintenance cost: High 4.单位成本光通量*: 2.5dollar/klm Unit cost of luminous flux*: 2.5 dollar/klm	1. 光效:中（10%~15%） Luminous efficiency: Medium (10%~15%) 2.平均寿命:5000小时。 Average life: 5,000 hours. 3.运维费:中 Operation and maintenance cost: Medium 4.单位成本光通量*:10dollar/klm Unit cost of luminous flux*: 10 dollar/klm	1.光效:高（>25 %） Luminous efficiency: High (>25%) 2.平均寿命:50000小时 Average life: 50,000 hours 3.运维费:低 Operation and maintenance cost: Low 4.单位成本光通量*: 16dollar/klm Unit cost of luminous flux*: 16 dollar/klm
对比结果 Conclusion	LED发光效率高，寿命长，运维费低，是未来发展趋势。 High luminous efficiency, long life and low operation and maintenance cost of LED make it the development trend in future.		

* 注：数据来源于美国能源部DOE（2014年）
* Note: The data is from the United States Department of Energy (2014).

对比之三：光品质比较
Difference 3: Quality of Light

	白炽灯 Incandescent lamp	荧光灯 Fluorescent lamp	LED LED lamp
图形比较 Photo			
特点比较 Features	1. 显色指数*Ra*:>90 Color rendering index Ra: >90 2.色温：不可变 Color temperature: Constant 3.光稳定性：较好 Stability: Quite good	1.显色指数*Ra*:>75~85 Color rendering index Ra: >75~85 2.色温：不可变 Color temperature: Constant 3.光稳定性：差 Stability: Poor	1.显色指数*Ra*:>70~90 Color rendering index Ra: 70~90 2.色温:可变（2700K~6500K） Color temperature: Variable (2700K~6500K) 3.光稳定性：较好 Stability: Quite good
对比结果 Conclusion	LED光源的色温可变，光稳定性好，应用广泛。 Due to the variable color temperature and high photostability of LED light source, it is widely used.		

对比之四：驱动控制比较
Difference 4: Drive control

	白炽灯 Incandescent lamp	荧光灯 Fluorescent lamp	LED LED lamp
图形比较 Photo			
特点比较 Features	1.启动时间：瞬间 Starting: Instant 2.达到稳态照明时间(70%)：瞬间 Duration to reach stable state (70%): Instant 3.驱动方式：无 Type of driving: N/A 4.驱动成本：无 Cost of driving: N/A	1.启动时间：瞬间 Starting: Instant 2.达到稳态照明的时间（70%）：5~15分钟 Duration to reach stable state (70%): 5-15 minutes 3.驱动方式：高压交流 Type of driving: High voltage AC 4.驱动成本：低 Cost of driving: Low	1.启动时间：0.1~0.5秒，但去电源化为发展方向 Starting: 0.1~0.5 second. But, driver-free LED is the trend. 2.达到稳态照明的时间（70%）：0.1~0.5秒 Duration to reach stable state (70%): 0.1~0.5 second 3.驱动方式：低压直流 Type of driving: Low voltage DC 4.驱动成本：高 Cost of driving: High
对比结果 Conclusion	LED光源稳态照明时间短，待解决驱动成本问题后，更具有竞争力。 With short steady–state lighting time, LED light source will be more competitive once drive cost is reduced.		

对比之五：设计及应用比较

Difference 5: Design and Application

	白炽灯 Incandescent lamp	荧光灯 Fluorescent lamp	LED LED lamp
图形比较 Photo			
特点比较 Features	1.光源与灯具：分体设计 Light source and lamp: Separate designs 2.光源组合：不可变 Combined light sources: Constant 3.光色变化：不可变 Change of colors: Constant	1.光源与灯具：分体设计 Light source and lamp: Separate designs 2.光源组合：不可变 Combined light sources: Constant 3.光色变化：不可变 Change of colors: Constant	1.光源与灯具：一体化设计 Light source and lamp: Integrated design 2.光源组合：灵活多变，可与建材一体化 Combined light sources: Variable, and can integrate with building materials 3.光色变化：全光谱RGB可变 Change of colors: Full spectrum RGB, variable
对比结果 Conclusion	LED可一体化设计、光源灵活多变、光色可变，应用为必然趋势。 Due to the possible integrated design, flexible and variable light source and variable light color of LED, it is inevitable to employ LED.		

对比之六：光源形状比较
Difference 6: Shapes of Photo Sources

	白炽灯 Incandescent lamp	荧光灯 Fluorescent lamp	LED LED lamp
图形比较 Photo			
特点比较 Features	1.光源形状：点光源 Shape of light source: Spot light source 2.光源设计：不可二次设计 Light source design: Secondary design unavailable 3.光源外壳材料：玻璃 Shell material: Glass	1.光源形状：线光源 Shape of light source: Linear light source 2.光源设计：不可二次设计 Light source design: Secondary design unavailable 3.光源外壳材料：玻璃 Shell material: Glass	1.光源形状：点、线、面光源 Shape of light source: Spot, linear and area light source 2.光源设计：可二次设计 Light source design: Secondary design available 3.光源外壳材料：非易碎性，任选透光材料 Shell material: Any non-fragile and light-transparent material
对比结果 Conclusion	因LED光源的形状可变性，透光材料任选性，并可二次设计，所以，LED广泛应用。 Due to the variable light source shape, optional light-transparent material and possible secondary design of LED, it is widely used.		

对比之七：可见光通信比较

Difference 7: Visible Light Communication

	白炽灯 Incandescent lamp	荧光灯 Fluorescent lamp	LED LED lamp
图形比较 Photo			
特点比较 Features	1.可见光通信：无 Visible light communication: N/A 2.信息承载力：低 Information bearer capacity: Low	1.可见光通信：无 Visible light communication: N/A 2.信息承载力：低 Information bearer capacity: Low	1.可见光通信：有 Visible light communication: Available 2.信息承载力：高，特定场所可替代wifi Information bearer capacity: High. It can replace wifi in certain situations.
对比结果 Conclusion	由于LED有可见光通信，信息承载能力高，实现可见光通信成为可能。 Because of the visible light communication and high information bearer capability of LED, it is possible to realize visible light communication.		

对比之八：光源的综合成本比较
Difference 8: Composite Cost of Light Sources

	白炽灯 Incandescent lamp	荧光灯 Fluorescent lamp	LED LED lamp
图形比较 Photo			
特点比较 Features	1.购置成本：低（800lm/60W/约3元） Acquisition cost: Low (800lm/60 watt/about 3 Yuan) 2.运维成本：高（每年换3次） Operation and maintenance cost: High (Three replacements per year) 3.能耗成本：高（60瓦，年耗电约175kW·h） Electricity bill: High (Annual power consumption will be about 175kW·h for a 60 watt lamp)	1.购置成本：中（800lm/15W/约12元） Acquisition cost: Medium (800lm/15 watt/about 12 Yuan) 2.运维成本：中（每年换1次） Operation and maintenance cost: Mediocre (One replacement per year) 3.能耗成本：中（15W，年耗电约44kW·h） Electricity bill: Mediocre (Annual power consumption will be about 44kW·h for a 15 watt lamp)	1.购置成本：高（800lm/9W/约25元） Acquisition cost: High (800lm/9 watt/about 25 Yuan) 2.运维成本：低（每年换0.2次） Operation and maintenance cost: Low (0.2 replacement per year) 3.能耗成本：低（9W，年耗电约26kW·h） Electricity bill: Low (Annual power consumption will be about 26kW·h for a 9 watt lamp)
对比结果 Conclusion	因为LED综合成本相对较低，所以LED应用是发展方向。 The low composite cost of LED makes it the development trend.		

对比之九：非可视照明应用比较
Difference 9: Nonvisual Lighting

	白炽灯 Incandescent lamp	荧光灯 Fluorescent lamp	LED LED lamp
图形比较 Photo			
特点比较 Features	1.健康医疗：无 Healthcare: N/A 2.种植养殖：少 Planting & breeding: A few 3.工艺用光：无 Technical lighting: N/A	1.健康医疗：少 Healthcare: A few 2.种植养殖：少 Planting & breeding: A few 3.工艺用光：少 Technical lighting: A few	1.健康医疗：多（心理、时差、睡眠、美容等） Healthcare: A lot (psychology, jet lag, sleep, beauty, etc.) 2.种植养殖：多（育种、补光、保鲜等） Planting & breeding: A lot (Seeding, supplemental light, fresh-keeping, etc.) 3.工艺用光：多（固化、光刻等） Technical lighting: A lot (Curing, photo etching, etc.)
对比结果 Conclusion	从LED的健康医疗、种植养殖、工艺用光三个方面的扩展应用上看，是未来的发展方向。 The extended applications of LED in healthcare, planting & breeding and technical lighting are going to be the development trends in future.		

LED照明技术的主要问题与对策

Main Problems & Countermeasures of LED Lighting Technology

5.1 LED照明技术的主要问题
Main Problems of LED Lighting Technology

序号 SN	主要问题 Main Problems	描述 Description
1	可靠性 Reliability	1. 驱动可靠性 Drive's reliability 2. 光衰和色温漂移 Luminous flux depreciation and color temperature shift 3. 灯具设计合理性 Reasonable lamp design 4. 大功率照明散热问题 Heat dissipation in large power lighting
2	光生物安全 Photobiological safety	1. 蓝光可能造成伤害 Possible blue hazard 2. 高亮度 High brightness
3	舒适性 Comfort	1. 眩光 Glare 2. 光斑 Light spot 3. 光品质 Quality of light
4	成本 Cost	初次购置成本偏高 The initial purchase costs are relatively high.
5	光效 Luminous efficiency	1. 光源光效有很大提升空间 Luminous efficacy of the light source can be largely improved. 2. 整灯效率仍有提升空间 Luminous efficacy of lamp can be improved.
6	标准及规范 Standards and norms	LED产品技术发展较快，相关设计、产品、检测标准有待完善和提高 LED technologies develop quickly. Related designs, products and test standards should be improved.
7	照明形态 Lighting forms	受传统照明产品影响，照明形态有待创新 As they are under the influences of traditional lightings, other forms of LED lightings should be innovated.
8	智能化 Intelligence	LED可控性特征没有完全得到发挥，智能化技术有很大应用空间 We have not exerted LED's features to control it. The intelligent technologies can be applied in much wider scope.

5.2 LED照明技术的主要对策
Countermeasures for LED Lighting Technology

序号 SN	主要对策 Counter-measures	描述 Description	备注 Remarks
1	政策的扶持 Policies Support	1. 完善现有补贴政策 Improve the existing subsidy policies; 2. 加大LED宣传力度 Put more efforts to make the people know about LEDs; 3. 扶持企业技术创新 Support companies to do more technical innovation.	
2	技术的提升 Technical advance	1. 提升上游装备的国产化水平，提高关键原材料的工艺水平 Locally produce more key upstream equipment, and increase processing quality for major raw materials. 2. 明确知识产权：规避专利风险 ，提高专利保护意识，开发具有自有知识产权的核心技术 Clarify intellectual properties: Avoid patent risks, increase the people's awareness of patent protection, and develop core technologies with intellectual properties of our own. 3. 加强LED产品模组化、标准化，提高产品兼容性、互换性 ，降低系统成本 Produce modulized standard LEDs and make the products more compatible and interchangeable so as to drop costs of the system. 4. 加强人才培养 Train more professionals of the industry.	
3	商业模式创新 Business mode innovation	1. LED关键技术、互联网技术和物联网技术结合进行跨界创新模式 Integrate key LED technologies, Internet technologies and Internet-of-Things technologies to create more modes. 2. 项目融资模式 Modes of financing 3. 主流大型产业联盟或者产业协同的模式 League of mainstream industries or industrial cooperation mode	

（续表）(Continued)

序号 SN	主要对策 Counter-measures	描述 Description	备注 Remarks
4	金融扶持 Financial support	1. 商业支付模式 Commercial payment mode 2. 金融支付模式 Financial payment mode	
5	LED标准的制定 LED standards development	1. 加快标准制定，提升行业门槛 Draft standards ASAP, and raise entry standard; 2. 加强标准体系建设和宣贯 Put more efforts in constructing, publicizing and implementing standard systems; 3. 严格市场监管 Strictly supervise the market.	

LED节能改造案例

LED-Based Energy-Saving Renovation Cases

案例之一：人民大会堂大礼堂LED节能改造案例

Case I: the Grand Auditorium of the Great Hall of the People LED Energy-saving Renovation

人民大会堂万人大礼堂照明改造为LED照明的分析 the Grand Auditorium of The Great Hall of the People LED Lighting Renovation Analysis				
项目概况 Project Overview	人民大会堂1959年建成投入使用，总建筑面积17.18万m²；2012年进行了办公楼LED照明节能改造。 In 1959, the Great Hall of the People was built up and put to use, with the overall floorage of 171.8 thousand m²; the office building of which was renovated into LED energy-saving lighting in 2012.			
改造地点 Renovation Objects	万人厅主席台 the Grand Auditorium rostrum	顶灯 Overhead light	观众厅 Auditorium	壁灯 Wall lamp
改造原因 Renovation Reasons	主席台上因灯光，温度过高，舒适性差 Lack of comfortability due to the high temperature of the original lighting on the rostrum	大栅灯泡寿命短，更换费时费力 Large grille lamps have short lifetime and change with effort	运营成本高 High operation costs	节能、省电 Energy-saving and power-efficient

（续表）（Continued）

人民大会堂万人大礼堂照明改造为LED照明的分析 The Grand Auditorium of The Great Hall of The People LED Lighting Renovation Analysis				
改造内容 Renovation content	更换为LED泛光灯 Change to LED floodlights	更换为LED吸顶灯 Change to LED ceiling lamps	更换为LED灯 Change to LED lamps	更换为LED直管荧光灯 Change to LED tube fluorescent lamps
改造效果 Renovation Effects	散热改善 Improved heat dissipation	节省人工 Save manpower	消除隐患 Hazard cleanup	节约运营费 Save operation costs
改造结论 Renovation Conclusion	（1）消除楼内灯具可能引起火灾的安全隐患，提高安全可靠性，节约运营成本。 Remove the fire hazard caused by inside lamps; improve safety and security and save operation costs. （2）主席台的LED投光灯更换后，照度提高1倍，解决了主席台温度过高问题。 The illuminance doubled after LED spotlights adopted on the rostrum and the problem of high tempreture on the rostrum solved. （3）会议厅LED灯管寿命长，更换方便，节约人工成本。 The LED lamps in conference rooms have long lifetime and can be easily changed, which would save manpower.			
产品厂商 Product Manufacturer	东莞勤上光电股份有限公司 Dongguan Kingsun Optoelectronic Co., Ltd. 广东朗视光电技术有限公司 Bright Vision Optoelectronic Technology Co., Ltd.			

案例之二：全国政协会议大厅LED节能改造案例

Case2: CPPCC Conference Hall LED Energy-saving Renovation

全国政协办公楼照明改造为LED照明的分析 CPPCC Office Building LED Lighting Renovation Analysis				
项目概况 Project Overview	全国政协办公楼1994年投入使用，总建筑面积 4.2万m²；2012年进行了办公楼LED照明节能改造。 In 1994, CPPCC office building was put to use, with the overall floorage of 42 thousand m²; it was renovated into LED energy-saving lighting in 2012.			
改造地点 Renovation Objects	常委会议厅主席台 The Rostrum in Standing Committee conference hall	常委会议厅 Standing Committee Conference hall	会议室 Conference rooms	地下车库 Underground parking
改造原因 Renovation Reasons	主席台上因灯光，温度过高，舒适性差； Lack of comfortability due to the high temperature of the original lighting on the rostrum .	大栅灯泡寿命短，更换费时费力。 Large grille lamps have short lifetime and change with effort.	牛眼射灯底座发热，易引起电气火灾等安全隐患。 Heat from the bases of bull-eye spotlights can cause safety hazards such as electrical fire.	运营成本过高。 High operation cost.

（续表）(Continued)

全国政协办公楼照明改造为LED照明的分析 CPPCC Office Building LED Lighting Renovation Analysis				
改造内容 Renovation content	更换为LED泛光灯 Change to LED floodlights	更换为LED吸顶灯 Change to LED ceiling lamps	更换为LED灯 Change to LED lamps	更换为LED直管荧光灯 Change to LED tube fluorescent lamps
改造效果 Renovation Effects	散热改善 Improved heat dissipation	节省人工 Save manpower	消除隐患 Hazard cleanup	节约运营费 Save operation costs
改造结论 Renovation Conclusion	（1）地下车库采用LED灯，半年时间节省的电费，收回一次投资。 Electric charge saved by LED lamps used in the underground parking in a half year can cover the initial investment. （2）消除楼内灯具可能引起火灾的安全隐患，提高安全可靠性，节约运营成本。 Remove the fire hazard caused by inside lamps; improve safety and security and save operation costs. （3）主席台的LED投光灯更换后，照度提高1倍，解决了主席台温度过高问题。 The illuminance doubled after LED spotlights adopted on the rostrum and the problem of high tempreture on the rostrum solved. （4）会议厅LED灯管寿命长，更换方便，节约人工成本。 The LED lamps in conference rooms have long lifetime and can be easily changed, which would save manpower.			
产品厂商 Product Manufacturer	广东朗视光电技术有限公司 Bright Vision Optoelectronic Technology Co., Ltd.			

案例之三：武汉神龙汽车厂LED节能改造案例

Case3: Wuhan Shenlong Automobile Factory LED Energy-saving Renovation

武汉神龙汽车厂改造LED照明项目分析 Wuhan Shenlong Automobile Factory LED Lighting Renovation Analysis			
项目概况 Project Overview	神龙汽车厂2009年投入使用，包括总装、焊装、冲压、涂装四个车间。2013年进行了LED照明节能改造。 Shenlong Automobile Factory came into use in 2009, including general assembly, welding, stamping and painting workshops, which were renovated into LED energy-saving lighting in 2013.		
改造地点 Renovation Objects	总装车间 General assembly workshop	焊装、冲压车间 Welding workshop, Stamping workshop	涂装车间 Painting workshop
改造原因 Renovation Reasons	总装车间灯具效率低，灯具本身耗电量大、灯具维护率高、光源衰减严重等问题，造成工厂地面照度严重不足。 In general assembly workshop, the Low efficiency, high power consumption, high maintenance and severe light attenuation of the former lamps make the ground illuminance woefully inadequate.	焊装车间存在大量的电火花，原有荧光灯管经常因电火花的冲击损坏，破裂，不但光效低，还带来安全隐患。 In the welding workshop, there are lots of electric sparks which impacted the original fluorescent lamps to be damaged and broken, which caused low luminous efficiency and safety hazards.	汽车车间带有很剧烈的振动，原有荧光灯具在振动环境下，存在灯管易松动，熄灭、甚至掉落的问题。 In painting workshop, there is strong vibration which made the original fluorescent lamps loosened, extinguished and even fallen.

（续表）（Continued）

武汉神龙汽车厂改造LED照明项目分析 Wuhan Shenlong Automobile Factory LED Lighting Renovation Analysis			
改造内容 Renovation content	更换为LED雷达感应格栅灯 Change to LED radar sensing grille lamps	更换为LED红外智能吊灯 Change to LED infrared intelligent droplight	更换为LED灯管配外置恒流电源 Change to LED tubes with external constant current power source
改造效果 Renovation Effects	散热改善 Improved heat dissipation	节省人工 Save manpower	消除隐患、节约运营费 Hazard cleanup, Save operation costs
改造结论 Renovation Conclusion	（1）二次节能方案，一次采用LED灯具替换荧光灯具节能达到50%。二次通过智能控制在原有基础上再节能50%。 Double energy-saving solutions: first energy-saving up to 50% by replacing fluorescent lamps with led lamps; second energy-saving up to 50% based on the previous result by intelligent control. （2）智能化控制，客户可以通过遥控器对灯具设置亮度等，满足客户的个性化需求。 Intelligent control, people can set lamp luminance by remote control to meet individual needs. （3）人体感应装置，灯具自动根据附近人员活动调节灯具亮度。 Human body sensors enable lamps to automatically adjust luminance in line with human activities nearby. （4）灯具采用无频闪技术，有效降低视觉疲劳，提高工作效率。 Use non-strobe technology, effectively relieve visual fatigue and enhance work efficiency. （5）项目总体节能率达到68%。 The overall energy saving rate of the renovation is 68%.		
产品厂商 Product Manufacturer	广东朗视光电技术有限公司 Bright Vision Optoelectronic Technology Co., Ltd.		

案例之四：全国政协文史馆LED节能案例

Case4: CPPCC Culture and History Museum LED Energy-saving Renovation

<table>
<tr><th colspan="5">全国政协文史馆LED照明项目分析
CPPCC Culture and History Museum LED Lighting Project Analysis</th></tr>
<tr><td>项目概况
Project Overview</td><td colspan="4">全国政协文史馆位于西城区金融街,于2012年建成投入使用，总建筑面积2万m²，地下4层、地上10层，该项目荣获“2012年度北京优秀照明工程”二等奖。
Located at Financial Street in Xicheng District, CPPCC Culture and History Museum was built up and came into use in 2012, the overall floorage of which is 20 thousand m² with 4 storeys underground and 10 storeys aboveground. The museum was awarded 2nd prize in Outstanding Lighting Project in Beijing 2012.</td></tr>
<tr><td>改造地点
Renovation Objects</td><td>地下停车场
Underground parking</td><td>楼梯
Stairs</td><td>通道
Passages</td><td>走廊
Aisles</td></tr>
<tr><td>改造原因
Renovation Reasons</td><td>长时间满负荷运转
能耗较高
Running at full capacity in a long time and high power consumption</td><td>非主照明区域
未实现按需照明
Lighting-in-need unimplemented outside the main lighting area</td><td>长时间开启
运营成本较高
Light always on
High operation cost</td><td>长明灯
镇流器热量高
Light always on, high temperature ballast</td></tr>
</table>

（续表）（Continued）

全国政协文史馆LED照明项目分析 CPPCC Culture and History Museum LED Lighting Project Analysis				
改造内容 Renovation content	使用为LED筒灯 Change to LED down lamps	使用为LED灯管 Change to LED tubes	使用为LED格栅灯 Change to LED grille lamps	使用LED智能通道灯 Change to LED intelligent aisle light
改造效果 Renovation Effects	散热改善 Improved heat dissipation	节省维护成本 Save maintenance costs	消除隐患 Hazard cleanup	节约运营费、按需照明 Save operation costs lighting in need
改造结论 Renovation Conclusion	(1)地下车库采用LED智能通道灯，半年时间节省的电费，收回一次投资。 Electric charge saved by LED intelligent aisle lamps used in the underground parking in a half year can cover the initial investment. （2）消除楼内灯具可能引起火灾的安全隐患，提高安全可靠性，节约运营成本。 Remove the fire hazard caused by inside lamps; improve safety and security and save operation costs. （3）LED灯具更换后，照度提高了1倍，且解决了原有灯具温度过高问题。 The illuminance doubled after LED lights adopted and the problem of high-tempreture lamps solved. （4）LED灯具寿命长，减少维护，节约人工成本。 The long lifetime and low maintenance of LED lamps can save labor cost.			
产品厂商 Product Manufacturer	广东朗视光电技术有限公司 Bright Vision Optoelectronic Technology Co., Ltd.			

案例之五：万达百货LED照明改造节能案例

Case5: Wanda Department Store LED Energy-saving Renovation

万达百货照明改造为LED照明的分析 Wanda Department Store LED Lighting Renovation Analysis	
项目概况 Project Overview	呼和浩特万达百货于2010年11月建成投入使用，商场1~5层经营面积2.8万m^2；2014年进行了百货大楼LED照明节能改造。 Hohhot Wanda Department Store was built up and came into use in November 2010. The business area on the 1st to 5th floor of the store amounts to 28 thousand square meters. The LED lighting renovation was made in 2014.
改造地点 Renovation Objects	商场卖场 Department Store
改造原因 Renovation Reasons	优化卖场室内商业环境，改善目前照明缺陷，节约用电 Better indoor shopping environment, remedy lighting defects and reduce power consumption.

（续表）（Continued）

万达百货照明改造为LED照明的分析 Wanda Department Store LED Lighting Renovation Analysis				
改造内容 Renovation content	更换为LED 筒灯 Change to LED down lamps	更换为LED 日光灯 Change to LED tubes	更换为LED 天花灯 Change to LED ceiling lamps	更换为LED 格栅灯 Change to LED grille lamps
改造效果 Renovation Effects	提高照明效果，改善室内灯具散热情况，节约用电 Better lighting effects, Improved indoor heat dissipation and reduced power consumption			
改造结论 Renovation Conclusion	（1）平均照度由100~200Lux提升至300Lux以上，平均照明效果提升一倍 Average illumination increases from 100-200Lux to mvaore than 300Lux and average lighting effects doubled. （2）节能率达到70%的高节能效果，既保证了整体照明效果，又降低了电费支出 Energy-saving rate reaches 70%. High energy-efficiency guarantees overall lighting effects while reduces electric charge.			
产品厂商 Product Manufacturer	广东昭信照明科技有限公司 Guangdong Real Faith Enterprise Group Co.,Ltd.			

案例之六：佛山南海区妇幼保健院LED照明改造项目

Case6: Foshan Nanhai District Maternity and Infant Hospital LED Lighting Renovation

<table>
<tr><th colspan="4">佛山南海区妇幼保健院改造为LED照明的分析
Foshan Nanhai District Maternity and Infant Hospital LED Lighting Renovation Analysis</th></tr>
<tr><td>项目概况
Project Overview</td><td colspan="3">佛山市南海区妇幼保健院是广州中医药大学附属南海妇产儿童医院，于1999年由南海区政府投资兴建，集保健、医疗、预防、科研、教学于一体，是三级甲等妇幼保健院。建筑面积约3万m^2。
Foshan Nanhai District Maternity and Infant Hospital is the maternity and children hospital in Nanhai District attached to Guangzhou University of Chinese Medicine. Invested by Nanhai district government and built in 1999, it is a third-grade class-A Maternity and Infant Hospital integrated of health care, medical treatment, prevention, scientific research and teaching. The floorage is around 30 thousand m^2.</td></tr>
<tr><td>改造地点
Renovation Objects</td><td>门诊楼
Outpatient building</td><td>住院部
Inpatient department</td><td>母婴大楼
Maternal and infant building</td></tr>
</table>

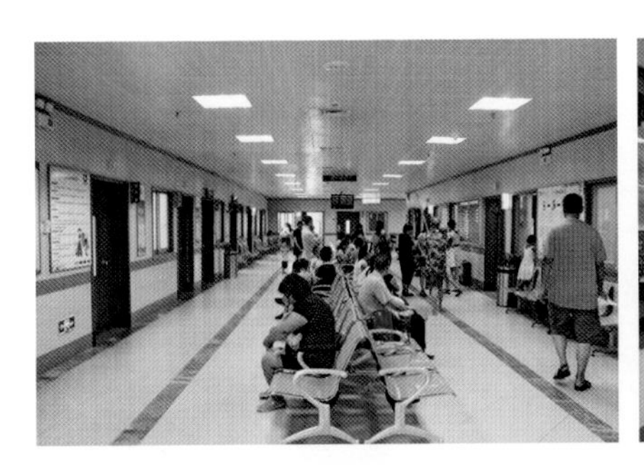

（续表）（Continued）

<table>
<tr><th colspan="4">佛山南海区妇幼保健院改造为LED照明的分析
Foshan Nanhai District Maternity and Infant Hospital LED Lighting Renovation Analysis</th></tr>
<tr><td>改造原因
Renovation Reasons</td><td colspan="3">优化就医环境，改善目前照明缺陷，节约用电。
Better medical environment, remedy lighting defects and reduce power consumption.</td></tr>
<tr><td>改造内容
Renovation content</td><td>更换为LED日光灯
Change to LED tubes</td><td>更换为LED筒灯
Change to LED down lamps</td><td>更换为LED吸顶灯
Change to LED ceiling lamps</td></tr>
<tr><td>改造效果
Renovation Effects</td><td colspan="3">节能率达到50%以上，节能效果显著。
Outstanding energy saving effects with energy saving rate above 50%.</td></tr>
<tr><td>改造结论
Renovation Conclusion</td><td colspan="3">更换LED灯后可使南海区妇幼保健院灯具总功率下降66.32%，年节约总费用约81.5万元。
After using LED, the total power of lamps in Nanhai District Maternity and Infant Hospital decreases by 66.32%, which saves costs amounting to 815 thousand Yuan per year.</td></tr>
<tr><td>产品厂商
Product Manufacturer</td><td colspan="3">广东昭信照明科技有限公司
Guangdong Real Faith Enterprise Group Co., Ltd.</td></tr>
</table>

案例之七：广晟国际大厦LED节能改造案例

Case7: Guangsheng International Tower LED Energy-saving Renovation

广晟国际大厦LED节能改造案例分析 Guangsheng International Tower LED Energy-saving Renovation Analysis			
项目概况 Project Overview	广晟国际大厦位于广州，面积约 16 万m^2，地上59层，高度约360m，为广州市第三高的建筑。 Project Overview: Located in Guangzhou and with the area of 160 thousand square meters, 59 storeys aboveground and the height of 360 meters, Guangsheng International Tower is the third tallest building in Guangzhou.		
改造地点 Renovation Objects	办公室格栅灯盘 Grille lamps in offices	会议室 Meeting rooms	地下车库 Underground parking
改造原因 Renovation Reasons	色温偏高，员工感受不舒服 High color temperature, which made staff uncomfortable	有眩光 Glare	运营成本高 High operation costs
改造内容 Renovation content	更换为LED平板灯 Change to LED panel lights	更换为LED筒 灯 Change to LED down lamps	更换为LED直管荧光灯 Change to LED tube fluorescent lamps
改造效果 Renovation Effects	办公室整体装修更为协调，照明环境改善 Lighting goes with the interior design in the offices. Better lighting environment	照明环境改善 Better lighting environment	节约运营费 Save operation costs

（续表）（Continued）

广晟国际大厦LED节能改造案例分析 Guangsheng International Tower LED Energy-saving Renovation Analysis	
改造结论 Renovation Conclusion	（1）本次改造综合节能率60.2%，节约运营成本 The overall energy-saving rate of the renovation is 60.2%. Save operation costs. （2）LED灯具寿命长，更换方便，节约人工成本 LED light has long lifetime and can be easily changed, which would save manpower. （3）照明环境改善，员工工作感受更为舒适。 Lighting environment improved, which has made staff more comfortable.
产品厂商 Product Manufacturer	广东广晟光电科技有限公司 Guangdong Ecorising Optoelectronic Science and Technology Co., Ltd.

案例之八：哈尔滨万达百货LED节能改造案例

Case8: Harbin Wanda Department Store LED Energy-saving Renovation

哈尔滨万达百货照明改造为LED照明的分析 Harbin Wanda Department Store LED Lighting Renovation Analysis			
项目概况 Project Overview	项目概况：哈尔滨万达百货是万达百货最早开业的百货店之一，面积约10万m^2。 Project Overview: Harbin Wanda Department Store is one of the earliest department stores under Wanda with the area around 100 thousand m^2		
改造地点 Renovation Objects	商场节能筒灯 Energy saving down light in stores	商场灯槽 Light troughs in Stores	仓库 Warehouse
改造原因 Renovation Reasons	亮度不足，耗电 Inadequate luminance while high power consumption	耗电量高 High power consumption	耗电量高 High power consumption
改造内容 Renovation content	更换为LED筒灯 Change to LED down lamps	更换为LED T5支架 Change to LED T5 bracket	更换为LED T8灯管 Change to LED T8 tube
改造效果 Renovation Effects	整体照度提高，购物环境更舒适，商品得到更好的展现 Enhanced overall illumination, better shopping environment and better display of goods	节约运营成本 Save operation costs	节约运营成本 Save operation costs
改造结论 Renovation Conclusion	（1）本次改造综合节能率65%，节约运营成本。 The overall energy-saving rate of the renovation is 65%. Save operation costs. （2）LED灯具寿命长，更换方便，节约人工成本。 LED light has long lifetime and can be easily changed, which would save manpower. （3）整体照度提高，购物环境更舒适，商品得到更好的展现。 Enhanced overall illumination, better shopping environment and better display of goods.		
产品厂商 Product Manufacturer	广东广晟光电科技有限公司 Guangdong Ecorising Optoelectronic Science and Technology Co., Ltd.		

附录
知名企业介绍

Appendix
Introduction to Renowned Enterprises

1. 中国建筑设计研究院（集团）

China Architecture Design & Research Group (CAG)

中国建筑设计研究院（又称“中国建筑设计集团”）是国资委直属的科技型中央企业，是中国城镇建设领域中业务覆盖面广、科技创新能力强的建设科技综合集团，始终位居中国建筑设计企业前列。

中国建筑设计集团发源于1952年创建的中央直属设计公司，2000年由原建设部直属的建设部建筑设计院、中国建筑技术研究院、中国市政工程华北设计研究院和建设部城市建设研究院组建而成。

China Architecture Design &Research Group (CAG) is a large technology-intensive enterprise directly administered by State-owned Assets Supervision and Administration Commission of the State Council (SASAC). It is a leading engineering and consulting company in China with a wide range of businesses and a strong capacity for technology innovation in urban and rural construction field.

CAG evolved from the previous Central Design Company founded in 1952, and was constituted as its current form in 2000 by consolidating several large and influential companies - Architectural Design Institute of the former Ministry of Construction (MOC), China

目前，中国建筑设计集团具有工程设计、城乡规划编制、文物保护工程、勘察设计、工程咨询、工程总承包等多项专业甲级资质；形成了城镇规划设计、民用建筑工程设计与咨询、基础设施与公共服务设施建设、建筑历史与文化遗产保护、节能环保、防灾减灾等技术领域的研究开发与技术服务的专业团队，成为中国建设科技领域的领跑者；在城镇规划、建筑设计、市政工程、标准规范、建设信息、工程咨询、室内装饰、风景园林等领域为中国乃至全世界的建筑及市政工程建设提供一体化的专业服务。

60年来，中国建筑设计集团先后设计完成了北京火车站、中国美术馆、国家图书馆、北京国际饭店、外交部办公楼、国家体育场（鸟巢）、首都博物馆、故宫保护、长城保护、引滦入津、西气东输、南水北调和长江三峡库区环境保护等标志性项目，“中国院”的品牌形象在国内外得到广泛认可。

Building Technology Development Center, North China Municipal Engineering Design &Research Institute and China Urban Construction Design &Research Institute of MOC.

At present, CAG holds various A-level professional qualifications for engineering design, urban and rural planning, cultural relic preservation, engineering survey and design, engineering consultation and EPC. CAG is respected as pioneer and leader in the sector for its strong R&D capacity and its outstanding professional teams in the field of urban planning and design, civil building engineering design and consultation, infrastructure and public service facilities construction, architectural historical and cultural heritage preservation, energy efficiency and environment protection, disaster prevention and reduction, etc. It aims to provide integrated services for architectural and municipal engineering construction in China and even abroad in the field of urban planning, architectural design, municipal engineering, code &standards development, construction information, engineering consultation, interior decoration and landscape architecture.

Over the past six decades, CAG has designed a number of renowned projects including Beijing Railway Station, National Art Museum of China, National Library of China, Beijing International Hotel, Office Building of the Ministry of Foreign Affairs, National Stadium (Bird's Nest), Capital Museum, the Forbidden City Preservation, the Great Wall Preservation, Luan River to Tianjin Project, West to East Gas Transmission, South-to-North Water Diversion Project, and the Environmental Protection of Three Gorges Reservoir Region. The brand of “CAG” has been widely recognized in China and abroad.

电话：（+86）010-57700800
邮箱：cag@cadg.cn
网址：http://www.cadreg.com.cn

Tel：（+86）010-57700800
Email：cag@cadg.cn
Web：http://www.cadreg.com.cn

2. 卓展工程顾问（北京）有限公司

China Team (Beijing) Co., Ltd.

卓展是一家独立的、跨行业的工程顾问公司，我们的服务覆盖建筑、工程等诸多领域，提供从咨询、设计到项目管理以及全方位工程顾问服务。

经过多年不懈的努力，卓展业已赢得广泛口碑以及客户的赏识，在提供创意和合理的解决方案方面一直处于行业的领先地位。客户不仅可以通过我们完成每一项任务和项目具体执行过程中所体现出来的专业技能、经验和高性价比的方案中获益，同时能够与各区域分公司的高级领导层及时沟通和协调。

China Team is an independent multi-disciplinary engineering firm providing a wide range of building and engineering services from basic Consultancy to complete Project & Construction Management and Total Solution Engineering.

Over the years, China Team has gain client's recognition and reputation, leading the way in creating and delivering better solutions. Our clients not only benefit from the technical experience and cost-effective solutions that we transfer to every assignment and project execution but also have immediate access to our top leadership stationed in our regional offices.

我们提供包括方案设计、初步设计、施工图设计、招投标、建筑监管、系统调试等全方位咨询服务。

我们承接的项目涉及多领域，从住宅、服务式公寓、酒店、写字楼到商业综合体，服务区域包括北京、上海、天津、重庆、辽宁、河北、河南、西安、四川、湖北、湖南、广东、海南等。

卓展正努力成为国际级的咨询公司，人才是最大的财富，我们将一如既往地追求并鼓励创新，营造最佳的工作氛围。

我们相信，创造并提供更好的解决方案，是满足客户需求的最佳方式。

We offer from-start-to-end consultancy ranging from conceptual design, preliminary design, detailed design, tendering, construction supervision, and testing & commissioning.

We have worked extensively across a broad spectrum of sectors from residential, serviced apartment, hotels, resorts, and office buildings, in regions such as Beijing, Shanghai, Tianjing, Chongqing, Liaoning, Hebei, Henan, Xi'an, Sichuang, Hubei, Hunan, Guangdong, Hainan, etc.

We strive for becoming a world-class consulting firm, as such we create working environment where initiative is encouraged and excellence is rewarded. People are our biggest asset.

We believe that Creating and delivering better solutions is the best way to address our clients' challenges.

电话：（+86）010-5900 0218
邮箱：info@china-team.com.cn
网址：http://www.china-team.com.cn

Tel.：（+86）010-5900 0218
Email：info@china-team.com.cn
Web：http://www.china-team.com.cn

3. 广东省半导体照明产业联合创新中心

Guangdong Solid State Lighting Industry Innovation Center

广东省半导体照明产业联合创新中心由广东省科技厅发起，由国家相关部门参与，省内科研机构、省内半导体照明上市企业、龙头企业共同出资成立，注册资本6300万，一期投资1.2亿元。联合创新中心主要面向产业链各个环节的创新需求，系统集成有效创新资源，完善创新服务功能，营造创新环境，建成广东LED产业发展战略智库、信息交

With the participation of the relevant departments of the state, Guangdong Solid State Lighting Industry Innovation Center （GSC） is sponsored by Guangdong Provincial Science and Technology Department, and established through the joint contributions of the provincial scientific research institutions, listed companies and leading enterprises in the semiconductor lighting industry. GSC has a registered capital of RMB 63 million, and the investment for the first phase is RMB 120 million.GSC mainly focuses

互枢纽、检测认证基地、技术创新桥梁、金融服务尖兵、人才培养高地、成果展示舞台。

联合创新中心以广东新光源产业基地为依托，通过与台湾工研院、香港应科院、国家半导体照明工程研发及产业联盟、清华大学、浙江大学等科研机构携手合作，汇集国际国内半导体照明巨头企业、科研团队和院所入驻中心，形成具有先进创新能力、创意服务能力和较强辐射带动能力的创新策源地，打造覆盖研发外包、检测检验、展示交易、教育培训、市场推广和金融支持等环节，面向广东、辐射全国、影响世界的LED产业创新服务集群。

on the innovation demands of various links of the LED industry chain, systematically integrates the effective innovation resources, improves the innovation service function, creates an innovation environment, and builds the development strategy think tank, information exchange hub, testing and certification base, technological innovation bridge, financial service vanguard, talent training highland, and achievement exhibition stage for the LED industry of Guangdong.

Relying on Guangdong New Light Source Industrial Base, the Center works together with Taiwan Industrial Technology Research Institute, Hong Kong Applied Science and Technology Research Institute, China Solid State Lighting Alliance, Tsinghua University, Zhejiang University, and other scientific research institutions, brings together international and domestic semiconductor lighting giants, scientific research teams and institutes to settle in the Center, forms a original place of innovation with advanced innovation ability, innovation service ability and strong radiation driving ability, creates the innovation service clusters in LED industry, which faces Guangdong Province, is provided with nationwide radiation, and will have an influence in the world, involving the links of R & D outsourcing, inspection and testing, exhibition and trade, education and training, marketing, and financial support, etc.

地址：佛山市南海区罗村新光源产业基地A8栋

网址：http://www.gscled.com
电话：0757-63860999
传真：0757-88362111
邮编：528226

Address：A8, Guangdong New Light Source Base, Luocun, Nanhai District, Foshan City, Guangdong Province, China
Web：http://www.gscled.com
Tel：0757-63860999
Fax：0757-88362111
P. C：528226

4. 广东广晟光电科技有限公司

Guangdong Rising Optoelectronic Technology Co. Ltd.

Ecorising是广东广晟光电科技有限公司（以下简称：广晟光电）旗下的LED照明品牌。广晟光电隶属于广东省广晟资产经营有限公司，是广东省最具规模、最具影响力的国有大型企业集团。广东广晟光电科技有限公司以“品质、诚信、创新、责任”为核心价值观，致力于培育“ecorising(广晟照明)”品牌，已在国内和14个国家进行商标注册。成员企业包括广东省中科宏微半导体设备有限公司、深圳中景科创光电科技有限公司、河南广晟高科技投资有限公司，设计MOCVD装备制造、LED封装、户外照明、室内照明等业务。广晟光电在地产、百货、商场、超市、办公等领域积累了丰富的行业经验

广晟光电始终坚持产品品质是企业的生命线，目前已通过ISO9000、ISO14000、OHSAS18000国际质量、环境和职业健康体系认证，全部生产运作流程都受ISO管控，对每一道生产检验工序都严格把关，确保产品质量的可靠性，主打产品已通过CCC、CQC、CE、广东省标杆体系、ROHS等国内、国际权威认证。

Ecorising is a LED lighting brand of Guangdong Rising Optoelectronic Technology, which affiliated to Guangdong Rising Assets Management Co., Ltd., one of the largest and most influential state-owned large enterprise group. Guangdong Rising Optoelectronic Technology Co., Ltd., takes “quality, integrity, innovation and responsibility” as core values. We have been committed to building the ecorising lighting brand and registrated trademark in China and 14 countries overseas. Our members include Zhongkehongwei Semiconductor Equipments Co., Ltd., Shenzhen China Cotrun Optoelectronic Technology Co., Ltd. and Henan Rising High-Tech Investment Co., Ltd. Our industry chain covers MOCVD equipment manufacturing, LED encapsulation, outdoor lighting and indoor lighting, etc. We have accumulated the wealth of industry experience in estate, department stores, shopping malls, supermarkets, offices and other fields.

Guangdong rising optoelectronic always adhere to the product quality is the lifeline of an enterprise. We have passed the international quality, environment and occupational health system certification of ISO9001, ISO14000, OHSAS18000. To ensure the reliability of the product quality, all production processes are under the control of ISO and every production inspection process is under rigorous management. Main products have passed CCC, CQC, CE, benchmarking system in Guangdong province, ROHS and other domestic and international authoritative certification.

电话：（+86）020-66263588
邮箱：gsled@ecorising.cn
网址：http://www.ecorising.cn

Tel.:（+86）020-66263588
Email：gsled@ecorising.cn
Web：http://www.ecorising.cn

广东精神
与你同行
HAPPY

5. 中科宏微半导体设备有限公司

Zhongkehongwei Semiconductor Equipments Co.

中科宏微公司是一家专门开展半导体照明产业核心装备MOCVD设计、制造以及氮化物LED外延应用的高技术公司，在MOCVD设备及相关LED外延的设计、开发和生产方面拥有一支实力雄厚的研发队伍，公司研发团队现有员工20余人，其中半数以上具有博士或高级职称，在半导体材料物理、外延工艺、机械真空、电气自动化等方面有丰富的研发经验。公司建有1500m^{2}的超净厂房、相关的配套设施以及高纯气体、软化水、高纯水等各种配套设施，购置了MOCVD设计制造以及外延材料测试的各种专用仪器和工具。

中科宏微公司是国内第一家提出交钥匙工程的国产MOCVD设备公司，也是国内第一家报道用自产设备生产的LED参数的公司。目前中科宏微公司生产制造的HiCRO®-Ⅰ型商用蓝光MOCVD已经销售5台，市场占有率居同类国产设备的领先位置。设备性能先后经中山大学、山东华光、扬州中科、晶科电子、湖南华磊、清华同方、台湾晶元光电等单位验证表明，用国产MOCVD设备外延的氮化镓材料及LED，其晶体质量、电学光学性能、均匀性等各项性能指标均达到较高水平。在氮化镓材料方面，其002摇摆曲线半高宽为184弧秒，102非对称衍射摇摆曲线半高宽为246弧秒；这些外延材料的常规技术指标，已经达到或超过国

MTM Semi is a hi-tech company specialized in development of semiconductor lighting industry core equipment MOCVD design, manufacturing, and nitride epitaxial LED application, of high technology company, which has a strong R & D team about MOCVD equipment design and related LED epitaxial development and production, The company has more that 20 R & D staffs, and above half of them have doctor or senior professional titles. They have rich experience about research and development in materials physics, semiconductor epitaxial process, mechanical, vacuum, electrical automation etc..This company has built 1500 square meters ultra clean plant and related supporting facilities, high purity gas, softened water, high purified water and so on, and has purchased some kinds of special instruments and tools which are often used for MOCVD design, manufacture and epitaxial material testing.

MTM Semi is the first domestic MOCVD maker which put forward the turn-key project and report LED parameters made by its own equipment. At present, MTM Semi had sold out five sets of HiCRO®- Ⅰ type commercial Blue ray MOCVD equipment. So the market share stands the leading position of domestic MOCVD makers. Equipment performance had been successively verified by Sun Yet-Sen University, Shandong Huaguang, Yangzhou Zhongke, APT Electronics, Hunan HuaLei Optoelectronic, Tsinghua Tongfang, and Taiwan EPISTAR so on. All tests show that in terms of Gallium Nitride materials and LED using domestic MOCVD equipment epitaxy the crystal quality, optical properties, electrical uniformity indicators have reached a higher level. In the gallium nitride material, its 002 rocking curve FWHM is 184 arc

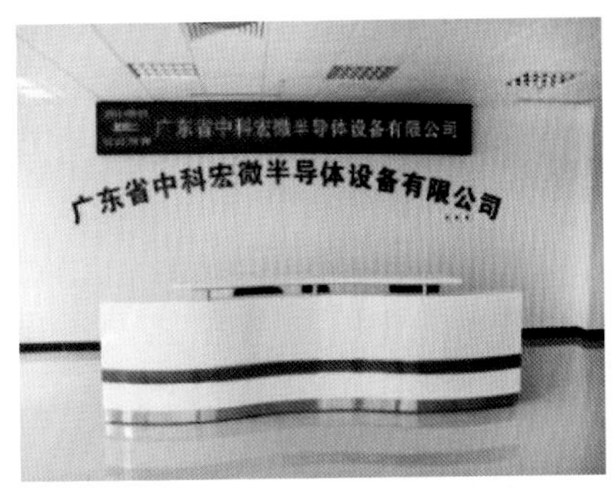

外同类商用设备的技术指标水平。在LED外延片方面，其单片片内波长均匀性偏差达到0.6%，全炉片间波长均匀性偏差达到1%以下，初步制作的氮化镓LED外延材料的小芯片发光功率达到17mW，制作的LED光效超过135lm/W以上。

物穷其理，宏微交替。中科宏微公司将努力搭建国际一流的MOCVD装备设计、制造和生产平台，形成年产MOCVD设备200台以上、年产值超过20亿元人民币的规模，成为国际一流的MOCVD高端装备制造企业和支撑我国半导体照明产业发展的重要企业。

and 102 asymmetric diffraction rocking curve FWHM is 246 arc. Those conventional technical indicators about epitaxial material have reached or exceeded similar foreign commercial equipment level. About LED epitaxial wafer, single piece of wave length uniformity deviation reaches 0.6%, the whole furnace inter slice wavelength uniformity deviation is less than 1%, the light power of preliminarily produced small chip reaches 17 mW, and LED optical efficiency exceeds 135lm/W.

MTM Semi will endeavor to build a world-class MOCVD equipment design, manufacturing and production platform, to form an annual output of MOCVD equipment exceeding 200, the annual output value of more than RMB 2 billion Yuan, to become a international first-class and most important MOCVD Equipment supplier, and to support China's semiconductor lighting industry development.

电话：（+86）020-66343760
邮箱：hsche@mtmsemi.com
网址：http://www.MTMsemi.cn

Tel.：（+86）020-66343760
Email：hsche@mtmsemi.com
Web: http://www.MTMsemi.cn

6. 广东昭信企业集团有限公司

Guangdong Real Faith Enterprise Group Co., Ltd.

广东昭信企业集团有限公司历练50载，从一家加工作坊小企业，成长为一家以光、电子产业为主体，集研发、生产、销售及进出口贸易为一体的现代企业集团，业务覆盖欧美、日本、东南亚等国家和地区，建有广东省技术中心、博士后工作站及省级工程实验室。

在“成为光、电子产业领先品牌企业”的目标追求下，昭信集团走进环保节能领域，与多家高等院校进行产、学、研合作，与行业结盟协同创新，已获得半导体LED领域国内外专利达150多项，发明专利30多项，目前，已形成半导体外延装备——外延芯片材料——封装——应用服务的LED全产业链。装备、封装、灯具、应用新技术，均获广东省企业自主创新纪录奖。

昭誉社会，信立未来，激励着昭信先后获取中国民营500强企业、广东省百强民营企业、广东省优秀民营企业、广东省自主创新标杆企业、广东省企业创新纪录金奖企业、LED最具竞争力品牌企业及全国先进基层党组织等荣誉称号。

Through its 50 years glorious history, Guangdong Real Faith Enterprise Group Co.,Ltd (ab. as Real Faith) grows from a small rattan ware processing workshop, into a modern group of enterprises with main business in lighting and electrics industry. Integrating R&D, manufacturing, sales and marketing,import and

export, Real Faith has covered its business at Europe, America, Japan, Southeast Asia and other areas, in addition, it has set up provincial technology center, Phd working station and provincial engineering laboratory.

In the pursuit of "Becoming a leading brand of lighting and electronics industry", Real Faith entered environment and energy saving field, worked with numerous of universities and institutes on manufacturing, academy and researching, allied into the industry for innovation. At present, Real Faith has owned over 150 patents in semi-conductor field, including over 30 invention patents, as well, covered the industrial chain from semi-conductor equipment,wafer, chip package to applications. Our technology of equipment, chip package, lighting fixture and lighting applications has won Guangdong Independent Innovation Record Award.

"Real to the society, Faith to the future". Inspired by this motto, Real Faith has won China Top 500 Private Enterprises, Guangdong Top 100 Private Enterprises, Guangdong Excellent Private Enterprises, Guangdong Enterprise Innovation Golden Award, the Most Competitive LED Brand, National Model Grass-roots Party Organization, and other honors.

7. 广东朗视光电技术有限公司

Bright Vision Optoelectronic Technology Co., Ltd.

广东朗视光电技术有限公司，致力于全球低碳、环保、节能事业，是一家集研发、生产、销售和服务于一体的LED综合应用产品与方案提供商，目前公司主要产品包括LED智能照明、LED办公照明、室内公共照明、LED地下空间照明和智能停车场。

Bright Vision Optoelectronic Technology Co., Ltd. devotes to the global low carbon, environmental protection and energy saving,it is a LED integrated application products and solutions provider with R&D, production, sales and service.Now our main products include Smart LED Lighting,Office LED Lighting,Indoor Public Lighting, Underground Space LED Lighting and Intelligent Parking Lot.

朗视光电跨学科整合智能控制技术、通信技术、互联网技术与LED照明技术，在行业率先提出“精益求精、注重品质”的产品理念，强调LED产品的功能性与适用性，通过创新性产品研发实现LED的真正应用。朗视光电在LED智能照明领域掌握多项世界核心控制技术，拥有多项发明专利，率先实现“按需照明”与“矢量照明”，成为全球LED智能照明技术最领先的企业。

公司先后完成了中国政协文史馆、全国政协主席楼、神龙汽车厂、海普瑞药业、上海世博园电力馆、新世纪环球中心、比亚迪汽车、北京石油管理干部学院、国家电网北京公司、北京理想国际大厦、西安科技资源中心、工商银行辽宁省分行、江苏省中医院、上海中山医院、南昌华南城、清华大学、华中农业大学、开元集团、西宁火车站、陕西省科技创业中心等项目，在国内外赢得了良好口碑。

BVLED interdisciplinary integrate intelligent control technology,communication technology, Internet technology and LED lighting technology . We first propose"keep improving, focus on quality" product concept in the industry, emphasize functionality and applicability of LED products and to realize LED's real application by innovative product development . BVLED master a number of core control technologies in the world on Smart LED Lighting field, and has a number of invention patents.We take the lead to achieve the "on-demand lighting " and " vector lighting ", thus to become the world's most leading intelligent LED lighting technology company.

Our company has accomplished the projects of C. P. P. C. C History Museum，National Committee Chairman Building, Dongfeng Peugeot & Citroen Automobile Plant，Hepalink, Shanghai World Expo Pavilion of Electricity，New Century Global Center, BYD Auto, CNPC, State Grid Company of Beijing, Beijing Ideal International Building, Xi 'an Science and Technology Resource Center, industrial and commercial bank Liaoning , Chinese medicine hospital of jiangsu, Shanghai zhongshan hospital, China South city, Tsinghua University, huazhong agricultural university, kaiyuan group, xining railway station, the shanxi provincial science and technology entrepreneurship center, etc, winning a good reputation at home and abroad.

电话：（+86）0755-33699550
邮箱：zc@bvled.net
网址：http://www.bvled.net

Tel.：（+86）0755-33699550
Email：zc@bvled.net
Web：http://www.bvled.net/